I0043786

Russia's Strategic Missile Carrier/Bomber Roadmap 2018-2040

PAK DA, Tu-160M2, Tu-95MSM & Tu-22M3M

HUGH HARKINS

Copyright © 2019 Hugh Harkins

All rights reserved.

ISBN: 1-903630-80-0
ISBN-13: 978-1-903630-80-8

Russia's Strategic Missile Carrier/Bomber Roadmap 2018-2040

PAK DA, Tu-160M2, Tu-95MSM & Tu-22M3M

© Hugh Harkins 2019

Centurion Publishing
United Kingdom

ISBN 10: 1-903630-80-0
ISBN 13: 978-1-903630-80-8

This volume first published in 2019

The Author is identified as the copyright holder of this work under sections 77 and 78 of the Copyright Designs and Patents Act 1988

Cover design © Centurion Publishing and KDP
Page layout, concept and design © Centurion Publishing

All rights reserved. No part of this publication may be reproduced, stored in a retrieval system, transmitted in any form, or by any means, electronic, mechanical or photocopied, recorded or otherwise, without the written permission of the publisher

The publisher and author would like to thank all organisations and services for their assistance and contributions in the preparation of this volume: Concern Radio Electronic Technologies, GosMKB Raduga, JSC Scientific Research Institute of Instrument-Development, V.V. Tikhomirov, JSC Myasishchev Design Bureau (EMZ), Kazan Aviation Factory n.a. S.P. Gorbunov, Ministry of Defence of the Russian Federation, NPO Saturn, OJSC Dubna Machine Building Plant, PJSC Beriev Aircraft, PJSC Tupolev, Rostec Corporation, S.P. Korolev Rocket and Space Corporation Energia, Tactical Missiles Corporation, The Boeing Company, Central Aerodynamic Institute, TsAGI, United Aircraft Corporation, UK Ministry of Defence, US Department of Defense

Citation guide: (Kret) Concern Radio Electronic Technologies, GosMKB Raduga, (Tikhomirov) JSC Scientific Research Institute of Instrument-Development, V.V. Tikhomirov, (EMZ) JSC Myasishchev Design Bureau (EMZ), Ministry of Defence of the Russian Federation, (DMB) OJSC Dubna Machine Building Plant, (Beriev), (NPO Saturn) NPO Saturn, PJSC Beriev Aircraft, (Tupolev) PJSC Tupolev, (Rostec) Rostec Corporation, (Energia) S.P. Korolev Rocket and Space Corporation Energia, (Boeing) The Boeing Company, (TsAGI) Central Aerodynamic Institute, (UAC) United Aircraft Corporation, (UK MoD) UK Ministry of Defence, (US DoD) US Department of Defense; (PSA) Post Session Analysis, session: 21 Aug 1987, 0823-09-27, (CIA) Project 9708, RDP96-00789R000300550003-6, , (Harkins) Harkins, H (2016) *Air War over Syria, Tu-160, Tu-95MS & Tu-22M3*, Centurion Publishing

CONTENTS

INTRODUCTION

In the second decade of the twenty first century the Russian Federation Long Range Aviation is undergoing a series of capability enhancements through the introduction of new long range strategic strike complexes, development of new long range strategic missile carries – PAK DA (Perspective Aviation Complex for Long Range Aviation) and Tupolev Tu-160M2 (new) – and a deep modernisation of all three Cold War legacy strategic missile carrier/long range bombers, resulting in the Tupolev Tu-95MSM, Tu-160M2 (modernised) and Tupolev Tu-22M3M. Whilst the volume is not intended to be a descriptive manual for the three Cold War legacy strategic missile carrier/bomber types – Tu-95MS, Tu-22M3 and Tu-160 – a brief description of the genesis, development and fielding of these weapon systems is provided for context.

The volume will draw on such information as is available from reputable sources to provide a picture of the Russian Federation strategic missile carrier/bomber roadmap as it stands in the second half of 2018. The evolution of the legacy Soviet era designs to reach the Tu-95MSM, Tu-160M2 and Tu-22M3M and associated weapons complexes intended for service into the 2040's and possibly beyond is detailed as is the new generation PAK DA. In regard to the PAK DA, information is naturally scarce as this program is shrouded in the highest levels of classification However, in 2018 significant data has been released to provide an informed outline of the complex, which is intended to constitute the backbone of Long Range Aviation toward the close of the 2020's.

All technical information concerning aircraft platforms and weapon complexes, along with the majority of the graphic material, has been furnished by the respective design bureau, manufacturers, experimental/research agencies and the Ministry of Defence of the Russian Federation.

1

THE COLD WAR LEGACY – TU-95MS, TU-160 & TU-22M3

In 2018, there were two new design/build airborne strategic missile carrier/bomber programs – the Tupolev Tu-160M2 and PAK DA (Perspective Aviation Complex for Long Range Aviation) – underway in the Russian Federation to equip Long Range Aviation in the third and fourth decades of the twenty first century and probably beyond. In addition, there are three further programs to enhance the capabilities and extend the service lives of the then extant strategic missile carrier/bomber fleets – Tupolev Tu-95MSM modernisation of the Tu-95MS strategic missile carrier, Tu-160M2 modernisation of the extant fleet of Tu-160 strategic missile carriers and the Tupolev Tu-22M3M modernisation of the Tu-22M3 missile carrier/bomber. Since the early 2010's the Cold War era armament of Kh-55 nuclear armed subsonic ALCM (Air Launched Cruise Missile), Kh-15 (nuclear armed) and Kh-22M series (conventional armed) supersonic ALCM and free fall unguided conventional and nuclear (it is unlikely that gravity nuclear bombs form a major element of Long Range Aviation armament, although they are nominally available for the Tu-95MS) have been augmented by the introduction of Kh-550(555) and Kh-101 (conventional armed) subsonic ALCM, Kh-102 (nuclear armed) subsonic ALCM, and conventional KAB guided bomb units.

In regard to PAK DA, the jewel in the crown of the Long Range Aviation capability enhancements, information is naturally scarce as this program is shrouded in the highest levels of classification. We know from UAC (United Aircraft Corporation) documentation that PJSC Tupolev is the prime contractor on the PAK DA project, certainly in regards to the airframe (UAC). However, any reference to PAK DA toward Tupolev will draw the inevitable PAK What? Such is the secrecy surrounding the program. That said, there are other avenues that can be investigated through information directly released by other entities, including UAC and the PAK DA end game customer, the MODRF (Ministry of Defence of the Russian Federation). Under 2018 planning it appears that the Tu-160M2 is being developed as a near term force augmenter and PAK DA is being developed as a longer term replacement for the Tu-22M3M and Tu-95MS bomber/strategic missile carriers.

Top: The trio of strategic missile carrier/long range bombers designs operating with Russian Long Range Aviation in 2018 – Tu-95MS leading a Tu-160 on the starboard wing and a Tu-22M3 on the port wing. Above: Tu-95MS operating with Long Range Aviation Tupolev/MODRF

As conceived, designed, built and deployed, the Tu-95MS, allocated the NATO reporting name 'Bear' H, would become the main element of the Soviet and later Russian air launched strategic nuclear deterrent armed with X-55 (Kh-55) air launched cruise missiles, which, like the carrier aircraft, entered service in the 1980's. Crown Copyright

Outwardly appearing like a metachronism, a Cold War relic of a bygone era, the Tu-95MS remains a highly capable intercontinental cruise missile carrying strategic strike aircraft. In 2018, the Tu-95MS forms the main component of Long Range Aviation's nuclear strike capability armed with Kh-55 and Kh-102 ALCM, with a secondary conventional role armed with conventional armed Kh-101 and Kh-550(555) ALCM. In terms of basic flight performance the turboprop powered Tu-95MS has a speed only some ~70 km/h less than that of the American jet powered Boeing B-52 strategic bomber that remains in service in 2018. In terms of operational airframes the B-52 fleet is considerably older than the Tu-95MS fleet, which was serial produced in the 1980's and early 1990's. The basic Tu-95 design, however, is as dated as that of the B-52 – the prototype Tu-95, '95/1', flying for the first time on 12 November 1952, the same year the YB-52 took to the air (Tupolev).

The basic design of the Tu-95 (remaining the case for the Tu-95MS) was a normal aerodynamic configuration featuring 'a high-lying cantilevered three-spar wing', which proved to be aerodynamically efficient when flying at high speeds, the Tupolev description reading 'This aerodynamic design provides high aerodynamic efficiency at high speed flight' (Tupolev). The high aspect ratio wing ('corresponding to the angle of its sweep and a set of profiles along the span') improves aircraft performance (Tupolev). The four x NK-12 turboprop engines, designed by N. Kuznetsov, drive AB-60K co-axial (contra-rotating) four bladed propellers (designed by K. Zhdanov). The engines, each of which generate 15,000 hp., are fed fuel housed in a central fuselage fuel tank and two central and four integral wing fuel tanks – refueling is centralised. In the Tu-95MS, range can be extended through in-flight refueling courtesy of a fixed in-flight refueling probe located on the upper nose section.

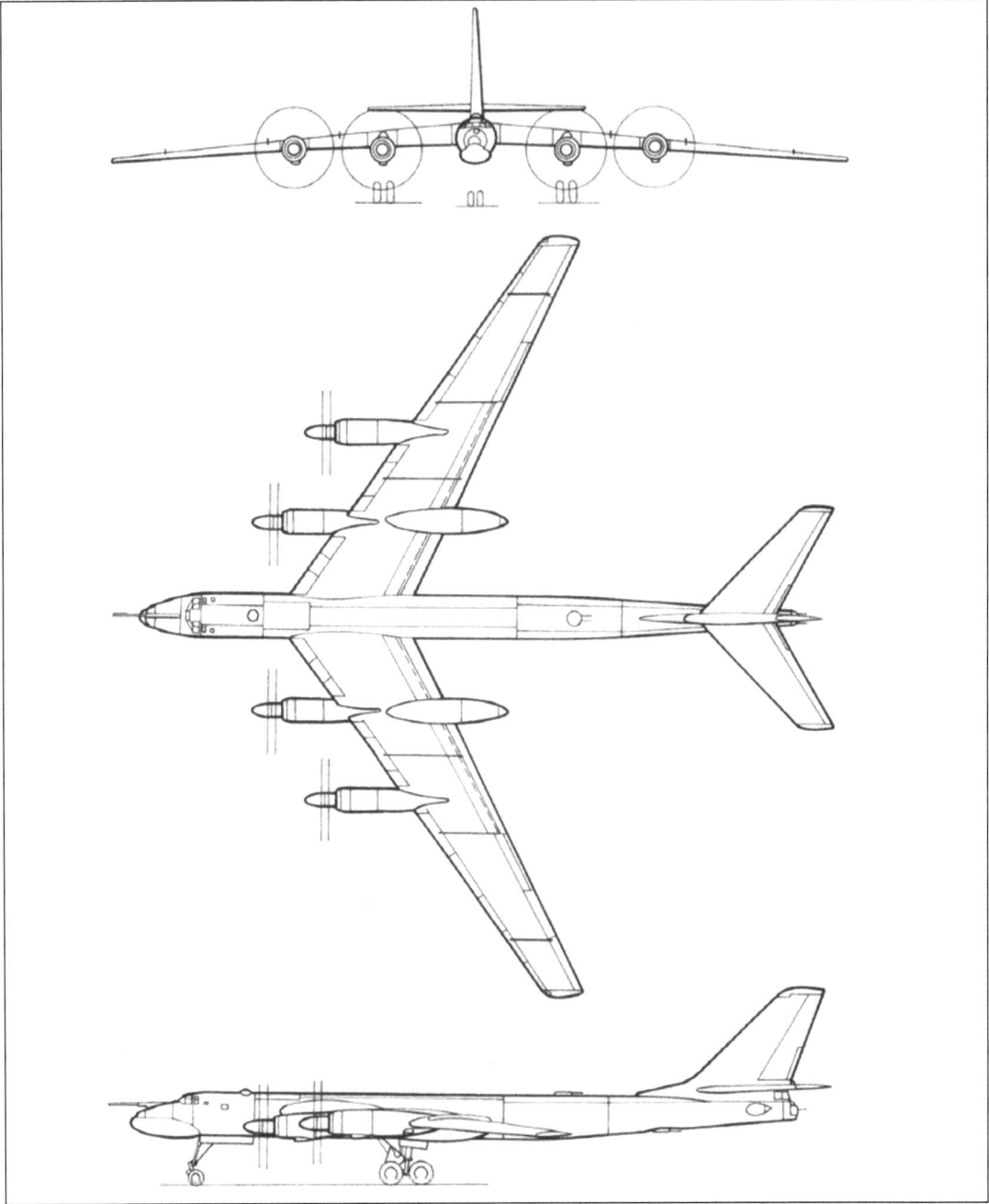

Three-view general arrangement drawing of the Tu-95MS strategic missile carrier. UAC

The Tupolev Tu-95MS Bear*

Strategic bomber and missile platform, one of the essential elements of the Russian strategic aviation forces

*NATO reporting name

UAC UNITED AIRCRAFT CORPORATION

Four NK-12M (MV) turboprop engines generating power at **4 x 15,000** h.p.

Some aircraft were named after Russian cities e.g. «Irkutsk», «Moscow» and so on

The only propeller-powered strategic bomber still in operational use today

Maximum speed **830** km/h, which makes it the fastest prop-driven aircraft

The maximum payload **20** tons

Wingspan (≈ **50** meters) is comparable with a 22-storey building in heigh

World record: **30 000** KM in 43 hours with 4 aerial refuelings over three oceans

Armament

- 2 x 23-mm twin-barrel cannons;
- 6 x X-55 long-range cruise missiles inside the fuselage (up to 10 missiles underneath the wings);
- Conventional and free fall bombs including nuclear.

Note

Special carrier aircraft Tu-95V was designed to test-drop the largest thermonuclear weapon ever - the Tsar Bomb.

Editor Olga Dubovitskaya
Designer Kirill Zadokhin

Ту-95МС

830 км/ч Скорость (макс.)

10,5 км Практический потолок

10 500 км Дальность полёта

7 чел. Экипаж

98,5 т Масса

Top: Graphic outlining a number of characteristic and historical facts about the Tu-95MS. This graphic details that conventional and nuclear gravity bombs are an armament option for the Tu-95MS. Above: Tu-95MS basic characteristic Russian language graphic – weight, 98.5 tons; operating altitude, 10.5 km; maximum speed, 830 km/h; range, 10500 km; crew, 7. UAC/MODRF

Tu-95MS of Russian Federation Long Range Aviation. UAC

Serial production of the first generation Tu-95 commenced in 1955 and the design entered Soviet service in 1957 – several hundred were manufactured in a plethora of variants – bomber, missile carrier, oceanic reconnaissance and special purpose. Many of the first generation variants served into the 1980's. Some Tu-95M's were converted to Tu-95U trainer configuration, serving into the early 1990's.

The Tu-142 anti-submarine warfare derivative, which conducted its maiden flight on 18 June 1968, was developed from the first generations Tu-95. This improved design would form the basis of the Tu-95MS strategic missile carrier as it was proposed that a variant of the Tu-142M, designated Tu-142MS (later designated Tu-95MS), would be developed as the new missile carrier. Work commenced in July 1977 and a converted Tu-142M conducted its maiden flight as the Tu-95MS prototype in September 1979. Tu-95MS serial production commenced at Taganrog aircraft production plant in 1981 before being transferred to Kuibyshev (now Samara) in 1983. Production continued into 1992, by which time the Soviet Union had been dissolved to form a Commonwealth of Independent States (Tupolev).

The Tu-95MS is equipped with a Leninets Obzor-MS long-range pulse-Doppler navigation and attack radar complex located in the aircraft nose section. This radar system, allocated the NATO (North Atlantic Treaty Organisation) reporting name 'Clam Pipe', has undergone updates that apparently include a synthetic aperture radar capability and Doppler beam sharpening modes for increased ground mapping resolution. As the Tu-95MS was designed to fly cruise missile attack profiles at medium altitudes there was no requirement for a terrain following radar capability (Harkins, 2016).

The Tu-95MS is powered by four x NK-12 turboprop engines driving four-blade contra-rotating propellers. MODRF

Top: A Tu-95MS strategic missile carrier trails Ilyushin Il-78M in-flight refueling tanker aircraft. Above: A Tu-95MS is prepared for flight in wintry conditions. MODRF/UAC

The self-defence and ECM (Electronic Counter Measures) suite included an SPS-171/172 electronically steerable phased-array jamming system, this being part of the overall Avtomat RWR (Radar Warning Receiver) complex. Other elements of the suite included an SPS-160 Geran active jammer, an Azovsky MAK-UT(L) L-082 IR (Infrared) MAWS (Missile Approach Warning System) and Avtomatika SPO-32/L-150 digital warning receiver, all of which are integrated with the Meteor-NM EWSP controller. The disposable element of the self-defence system consisted of an APP-50 chaff/flare dispenser system. The hard kill element of the self-defence suite consisted of a two GSh-23 23 mm twin barrel cannon complex housed in the tail position (Harkins, 2016).

For its primary nuclear strike role the Tu-95MS is armed with the Kh-55 or Kh-102 air launched cruise missile, six of which are carried on and launched from a SRPE MKU-5-6 rotary launcher located in the internal weapons bay. While six is the maximum number of Kh-55 missiles that can be carried by the Tu-95MS-6, a further ten such missiles can be carried on under wing stations on the Tu-95MS-16 (UAC).

Tu-95MS-16. MODRF

Referred to in Russian Federation Long Range Aviation service as an Intercontinental Strategic Multi-Mode Missile Carrier designed to strike high value installations on intercontinental deep penetration missions, the Tu-160 (allocated the NATO reporting name 'Blackjack') possesses enormous combat potential in both nuclear and conventional long-range strike scenarios armed primarily with air launched cruise missiles.

Подготовленный к полёту опытный Ту-160 («270-01»), г. Жуковский, 08.04.1982

Page 10-11: Series of three images showing the prototype Tu-160 during a development flight on 08.04.1982. UAC

Possessing an unquestionable outward resemblance to the American Rockwell (later Boeing) B-1 strategic bomber, the Tu-160, the heaviest combat aircraft ever built, is not only larger than the B-1, but possess greater overall performance, particularly in such areas as kinetics, maximum speed being around 400 km/h in excess of that of the B-1B. The Tu-160 ceiling is also considerably greater than that of the American aircraft, these values being 14000 m and 9140 m respectively (Tupolev and Boeing values). Unrefueled range of the Tu-160, stated as ~14000 km, is also considerably greater than that of the B-1B. Comparisons with the B-1 are dismissed among elements of the Tu-160 design team. The Russian aircraft, as noted above, being much larger and possessing very different performance parameters.

PJSC Tupolev stated that the Tu-160 was more advanced and ultimately more capable than the American aircraft. Certainly in regards to airframe and flight performance this would ring true. However, the B-1B has benefited from huge sums being spent on avionics and weapon capability updates, most recently the incorporation of the B-1B integrated battle station, making it probably the most versatile of the US bomber triad, with a diversity of weapons able to be employed. That said, the Tu-160, with its 4500 km range Kh-101/102 cruise missiles, possesses a stand-off strike capability that cannot be rivalled by any other aircraft, except perhaps the turboprop powered Tu-95MS described above, which in its Tu-95MS-16 form can more Kh-101/102 missiles than the Tu-160. The introduction of Tu-160M2 will increase combat effectiveness of the Tu-160 design by a factor of 2.5.

Known affectionately in Russian service as the 'White Swan', the Tu-160, allocated the NATO reporting name 'Blackjack', represents the pinnacle of current (2018) Russian long range airborne strike capability. Tupolev

Design work to meet the requirement that would lead to the Tu-160 commenced following a decree of the Soviet government, dated 28 November 1967, which called for development of a long-range missile carrier/bomber aircraft. Tupolev involvement in the late 1960's was merely as an observer party. It was not until 1970, following a capability and capacity assessment that Tupolev committed to development of a new strategic missile carrier/bomber, this task being allocated to the bureau's 'K' department (Tupolev). Competition to Tupolev was present from other design bureaus.

Design work on the new Soviet supersonic bomber commenced at Tupolev Design Bureau in the early 1970's under the design leadership of Valentin Ivanovich Bliznyuk, but under the general supervision of A.A. Tupolev (Tupolev). The basic concept was formulated by 1972, although much work was required in many areas, including materials of manufacture, related to operational parameters, and the choice of power plant – design of the aircraft involved around 800 different bureau and organisations (Tupolev). The basic design was settled by the middle of the decade and a preliminary design for the full size aircraft was prepared in 1976-1977. This led to commencement of production of the three prototype/development aircraft in cooperation with at the Kazan aircraft manufacturing plant (Tupolev). The final assembly of the prototype Tu-160 was completed at Zhukovsky near Moscow, in January 1981 and the aircraft underwent a series of ground testing. On 14 November 1981, the aircraft, with pilot B.I. Veremei at the controls, conducted a series of taxiing tests in preparation for its 30 minute maiden flight, which was conducted on 18 December that year (Tupolev).

Tu-160 RF-94109 serving with Russian Federation Long Range Aviation in the twenty first century. MODRF

The Tu-160 design is described by Tupolev as being built on 'a low faired scheme with a variable [variable-geometry] swept back wing, tricycle landing gear, all moving stabilizer fin', and all moving horizontal stabilizers (Tupolev). The control surfaces on the variable-geometry wings include slats, double-slotted flaps (for roll control and guidance) and drooped ailerons. As the wings are swept backwards boundary layer fences are raised from the inner section of the variable sweep wings improving lift and drag ratios, particularly in the fully swept back, 65°, position. The airframe is constructed mainly of titanium, steel alloys, heat-treated aluminium alloys and composite materials (Tupolev).

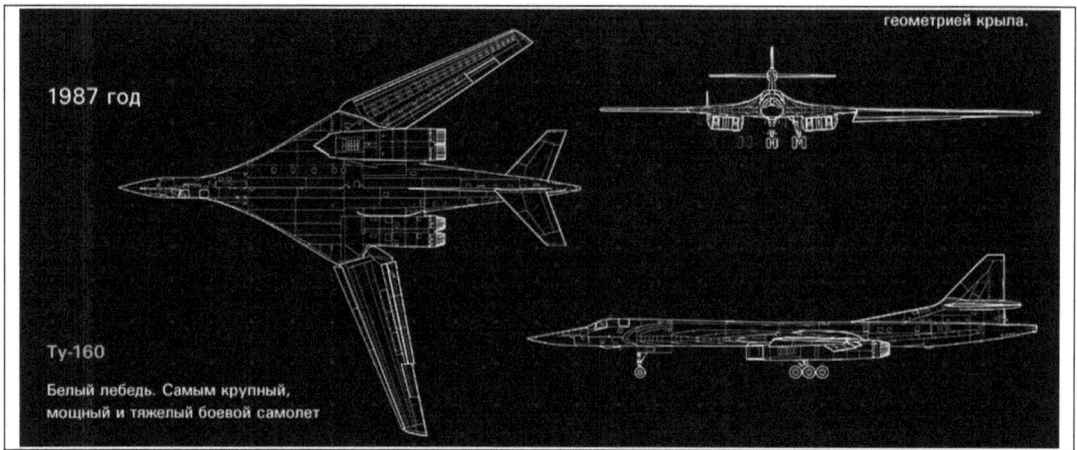

геометрией крыла.

1987 год

Ту-160

Белый лебедь. Самым крупный,
мощный и тяжелый боевой самолет

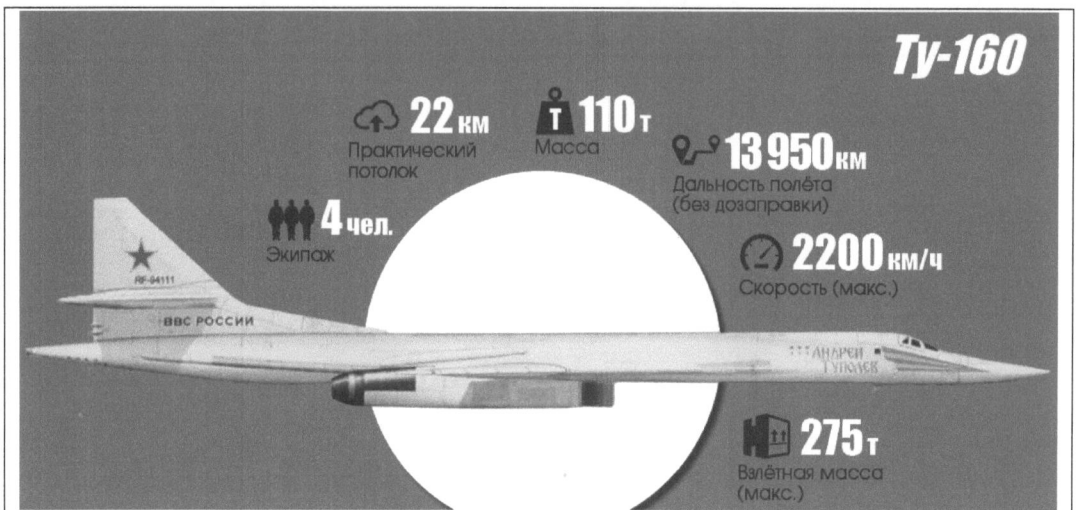

Ту-160

⌂ 22 км
Практический
потолок

⚖ 110 т
Масса

⌖ 13 950 км
Дальность полёта
(без дозаправки)

👥 4 чел.
Экипаж

⊙ 2200 км/ч
Скорость (макс.)

275 т
Взлётная масса
(макс.)

Top: Three-view general arrangement drawing of the Tu-160. This Russian language graphic depicts the aircraft with its port wing in the forward (upswept) position and, on the starboard side (facing forward to the aircraft fore section) wing (this is an underside view of the port wing) in the fully swept back position. Above: Russian language graphic showing the basic characteristics of the Tu-160 – crew, four; operating ceiling, 22 km; basic weight, 110 tons; range, 13950 km; maximum speed, 2200 km/h and maximum take-off weight, 275 tons. Kret/MODRF

The Tu-160 is powered by four x NK-32 turbofan engines grouped in two's in nacelles located in two pods mounted on the underside of the fixed wing sanctions. These units, with a thrust rating of 25 tons (~22680 kg) in afterburner – combined thrust of 100 tons = more than 90700 kgf – being the most powerful engines ever fitted to a combat aircraft. Variable geometry ramps in the engine intakes optimise airflow to the engines during various flight parameters, bestowing upon the Tu-160 its 1800 km/h flight speed (Tupolev value). A stand-alone TA-12 auxiliary power unit provides power to aircraft systems independent of the engines while the aircraft is on the ground.

The Tu-160 is powered by four x NK-32 afterburning turbofan engines, providing a combined thrust in excess of 90700 kgf for take-off. UAC

Самый мощный в истории военной
авиации сверхзвуковой самолет,
имеющий наибольшую среди
бомбардировщиков максимальную
взлетную массу

Стратегический
бомбардировщик-
ракетоносец
ТУ-160

Top: Russian language Tu-160 graphic depicting the basic variable-geometry layout of the Tu-160. The text basically translates to highlight that the Tu-160 strategic missile (rocket) carrier is the most powerful and largest mass combat aircraft in history. Above: On touch down during the landing process, three large braking parachutes are deployed from the rear nose cone. MODRF/UAC

Fuel tanks feeding the engines include two front fuselage/fixed wing section blend and one in rear fuselage, the three holding a combined total of 171 tons (various sources state conflicting values, 171 tons being the upper value and 130 tons being the lower value) of fuel, bestowing upon the aircraft a very high unrefueled flight range. The already impressive range can be further increased by in-flight refueling courtesy of a retractable in-flight refueling probe mounted on the aircraft upper nose section.

A Tu-160 trails behind an Il-78M in-flight refueling tanker aircraft. MODRF

The aircraft interior accommodates a forward fuselage crew compartment with pilot and co-pilot seated side-by-side at the front and the navigator and defensive systems operator side by side behind, all crew being seated on K-36D zero zero ejection seats. Designed from the outset for extremely long-range flights, a toilet facility, kitchen and sleeping berths are provided for the crew.

The primary targeting and navigation sensors include the nose mounted Leninets Obzor-K radar system, a modified variant of the Obzor-MS fitted in the Tu-95MS, which is integrated with a Sopka terrain following radar, allowing the Tu-160 to fly low-level penetration missions if required. The Obzor-K detection range is around 300 km, the range of the Obzor-MS fitted in the Tu-95MS being only marginally different, if at all. The Groza OBP-15T remote electro-optic TV (Television) sighting system, located in a ventral position on the forward fuselage, is designed for the employment of gravity unguided munitions. As well as radar and bomb-sighting systems the targeting complexes can include a SMKRITTs RORSAT, or equivalent, targeting datalink receiver (Molniya satcom) and an AFA-15 strike camera. The navigation suite includes a satellite navigation system, originally an early system that predated the Russian GLONASS (Globanaya Navigozionnaya Sputnikovaya Sistema

(Global Navigation Satellite System)) and US NAVSTAR (Navigation Satellite Timing and Ranging) systems. The current system receives its data from the GLONASS constellation. Other elements of the navigation and communications suite are thought to be similar to items of the Tu-22M3 suite – NK-45 navigation system, DISS-7 Doppler Navigation system, RV-5 low altitude altimeter, RV-18G radio altimeter, RSBN-PKV TACAN, R-832M UHF (Ultra High Frequency) and R-847 HF (High Frequency) systems. Appropriate navigation elements are integrated with the autopilot.

Toward the rear nose cone are elements of the self-defence suite designed for the detection and jamming of threat radar systems – the defensive systems may include an AG-65 ECM automatic noise generator, Avtomat-2 and 3 RWR for the detection of airborne radars, such as those carried by fighter aircraft, and ground based radar systems respectively. At the extreme point of the tail cone is located the thermal detection system designed to detect threats such as threat missiles and aircraft in the rear hemisphere. This system is thought to be along the lines of the L-082 MAK-UT(L) IR MAWS equipping the Tu-95MS. The disposable element of the self-defence suite consists of the APP-50 chaff/flare dispenser system, flare ejection bays being located on the rear fuselage lateral/undersides.

There are two large internal weapons bays, the doors for which open downward and outward from the fuselage underside. The bays can accommodate 22 tons of free fall munitions or a six round rotary launcher in each bay for an operational load of 12 x Kh-55SM/Kh-102 nuclear armed cruise missiles or Kh-101 conventional armed cruise missiles. The Kh-55SM has a flight range in excess of 2000 km while the Kh-102/Kh-101 have a flight range up to 4500 km.

The two large internal weapon bays can each accommodate a six round rotary launcher for the deployment of conventional or nuclear armed cruise missiles. Kret

The prototype Tu-160. UAC.

The Tu-160 prototype attained supersonic speed in February 1982. Serial production commenced a few years later and the design entered operational testing in October 1984. The first operational Soviet Tu-160 unit was the 184th Bomber Regiment of the 37th Air Fleet, based at Priluky air base, Poltava Oblast, Soviet Ukraine, the first two aircraft arriving there on 17 April 1987 (Tupolev). The primary combat scenario envisioned the Tu-160's employing Kh-55 nuclear armed ALCM to strike North American strategic targets (Tupolev). Initial planning called for Long Range Aviation to field up to 100 Tu-160's. When the Soviet Union was dissolved on 25 December 1991 production totaled only 35 airframes, some of which were incomplete. Production continued at drip pace and, with the passing over into a new century, a total of 36 Tu-160's had been built (this value apparently including a single new-built aircraft delivered in the early 2000's). The above values belied the numbers of Tu-160's available for operations with the Russian Federation, which inherited Long Range Aviation, as 19 of the 23 or so operational Tu-160 aircraft were inherited by Ukraine on the break-up of the Soviet Union in 1991. The Tu-160 aircraft remaining in Russia were allocated to the 121st Guards Heavy Bomber Regiment at Engels air base, Saratov, the six such aircraft being on strength by 1994 being insufficient to constitute a viable strike force. Ten of the Ukrainian Tu-160's were scrapped, one went to a museum, and the remaining eight were purchased by Russia in 1999, along with three Tu-95MS-16 and several hundred Kh-55 cruise missiles, under a contract valued at US $285 million, which was deducted from Ukraine's natural gas debt to Russia. The first ex-Ukrainian aircraft arrived at Engels on 6 November 1999, followed by the remaining seven over the next several months. In 2001, air to air refueling training for the Russian Tu-160 fleet commenced. Over the next few years the fleet worked up, being officially declared operational in Russian Federation Air Force Long Range Aviation in 2005. The sixteen Russian Tu-160's constituted a viable strike force, with several employed for training or undergoing maintenance at any particular time (Harkins, 2016).

The Tu-160 established 44 speed and altitude word records in two phases of flights in October 1989 and May 1998 respectively (Tupolev). Many long-range flights have been conducted. Two Tu-160's, 'Alexander Molodchiy' and Vasily Senko', were flown from Engels to Libertador, Venezuela in 2008, the first time a Tu-160 of the Russian Federation had operated from a foreign air base. A further record was set for range in June 2010 when two Tu-160's, employing two in-flight refuelling, flew 18000 kilometers in a time of 24 hours and 24 minutes (Tupolev).

Tu-160's of Russian Federation Long Range Aviation. MODRF

Although sharing the Tu-22 designation the Tu-22M (above) had little in-common with the first generation Tu-22 (top). The early variants of the Tu-22M included the Tu-22M2, NATO reporting name 'Backfire' B, an example here being shadowed by a USN Grumman F-14A Tomcat fleet air defence fighter. NPO Saturn/US DoD

Tu-22M3 of Russian Long Range Aviation. Tupolev

The smallest and shortest legged of the Russian missile carrier/long range bomber triad, the Tupolev Tu-22M3 (NATO reporting name 'Backfire' C) was designed as a supersonic strike bomber capable of attacking land and seaborne targets with free fall bombs and guided missiles. In regards to the latter, the Tu-22M3 took on the role of aircraft carrier killer, tasked with destroying NATO aircraft carrier battle groups in the event that the Cold War turned hot.

Although the designation would imply that the Tu-22M was a further development of the Tu-22, this is illusory as the design, as conceived, developed and built, was completely new. The Tu-22 designation was a necessity to court funding for the program in a time when intercontinental and intermediate range ballistic missiles held favour within the Soviet political leadership over intermediate range bomber aircraft – updates of existing designs appearing more affordable than new designs. Designed as a Tupolev Tu-16 replacement, the first generation Tu-22, '105 Project', was officially born on 15 August 1955, the prototype conducting its maiden flight on 21 June 1958. From 1959 until 1969, 311 Tu-22's were built in a number of variants, including Tu-22A and Tu-22B bombers, Tu-22P Scout (reconnaissance), Tu-22K missile carrier and Tu-22U trainer.

When design work commenced in 1965, without state funding, the Tu-22M (Modernised), initially known under the product code '145' and official designations of 'AM' and 'YN', was touted as a profound modernisation of the Tu-22K (Tupolev). As the design progressed the final product code '45' was applied and by 1967, with the engines mounted in the rear fuselage with the trunks on both sides of the aircraft keel, combined with the adoption of variable-geometry (swing-wings), any resemblance to the Tu-22K faded. A resolution of the Soviet government authorising a modification from the Tu-22K was issued on 28 November 1967. This design, which was to be powered by two NK-144-22 engines and referred to as the Tu-22KM. would evolve into the Tu-22M powered by two 25000 kgf NK-25 afterburning turbofan engines housed side-by-side in the rear fuselage (Tupolev). Variable-geometry air intakes feed air to the engines through the above mentioned lateral fuselage mounted intake trunks.

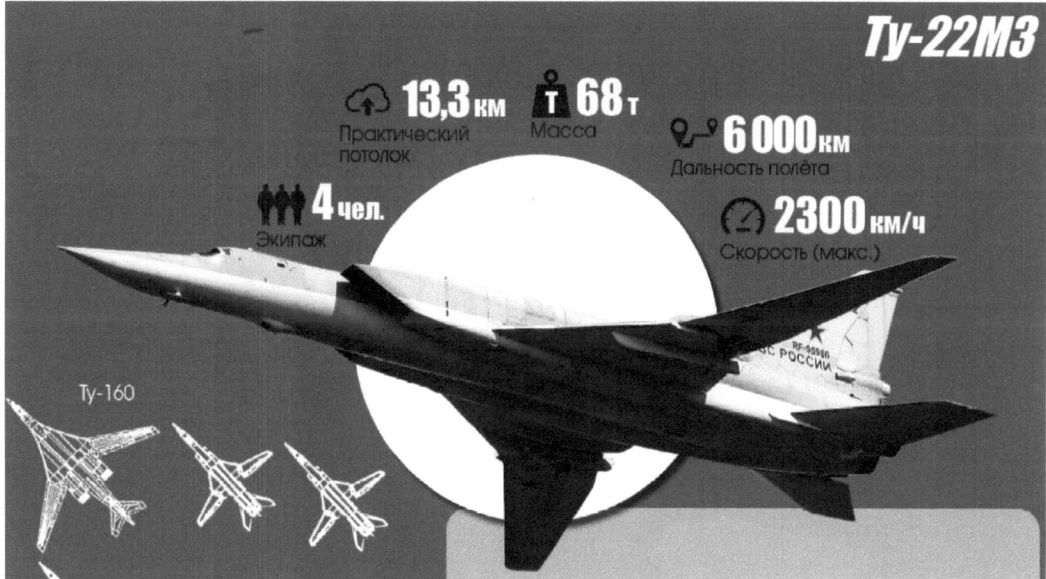

Russian language graphic showing the basic characteristics of the Tu-22M3 – maximum speed, 2300 km/h; ferry range, 6000 km; weight, 68 tons; operational ceiling, 13.3 km and crew, 4. MODRF

As designed, much of the Tu-22M flight and weapon deployment operations were automated. The aircraft was designed along the lines of a normal aerodynamic layout with a cantilever low set variable-geometry wing configuration featuring a semi-monocoque fuselage that rested on a tricycle undercarriage, the six-wheel main units of which retract to lie in the fixed inner wing section at the fuselage wing join. The twin-wheel nose unit retracts aft to lie in the forward fuselage underside. The airframe is built from mainly aluminium and steel alloys and some magnesium. The wing consists of a fixed inner part and the variable-geometry section, which can sweep from 65° to 20°. Wing control surfaces consist of the flap system, including slats, three section double slotted flaps and three section spoilers, without ailerons. For roll control the spoilers are operated differentially and also double as airbrakes, the function of lateral control being preserved. The design incorporated a single vertical tail fin with all-moving stabiliser. For ground operations power is supplied by an AP TA-6A auxiliary power unit located in the dorsal fin (Tupolev).

The Tu-22M3 is equipped with a Leninets PNA-D attack radar – earlier variants of the Tu-22M were equipped with the PNA-A/B variants. The PNA-D complex can apparently perform the function of terrain following and Doppler beam sharpening ground mapping modes. A Groza OBP-15T remote TV sighting system, located in a ventral position on the forward fuselage, is designed for the employment of unguided gravity munitions. This system was carried over to the Tu-160, although visually the optics windows appear distinct from each other. The targeting system also included a SMKRITTs RORSAT (Radar Ocean Reconnaissance Satellite) targeting datalink receiver (Molniya satcom) and an AFA-15 strike camera. The communications and navigation suite, the latter integrated with the ABSU-145M autopilot, consists of the NK-45 navigation system, DISS-7 Doppler Navigation system, RV-5 low altitude altimeter, RV-18G radio altimeter, RSBN-PKV TACAN, R-832M UHF and R-847 HF communications systems. The defensive systems include an AG-65 ECM automatic noise generator, Avtomat-2 and 3 for the detection of radar systems and an L-082 MAK-UL IR MAWS. The disposable element of the self-defence suite consists of the APP-50 chaff/flare dispenser system while a hard-kill self-defence capability is provided by the GSh-23 23 mm twin barrel cannon, with 1200 rounds of ammunition, housed in a barbette in the tail. The cannon complex incorporates an upgraded PRS-4KM Kripton fire control and ranging/tail warning radar integrated with a TP-1 tail warning/fire control TV camera system.

The Tu-22M3 has an impressive performance for an aircraft of its weight class, 68 tons normal and ~124 tons maximum take-off. The design features a maximum speed of 2300 km/h (MODRF value), an operational ceiling of 14000 m and an unrefueled operational tactical range of 2200 km (MODRF). Range can be increased by the application of in-flight refueling courtesy of the retractable in-fight refueling probe located on the upper nose section.

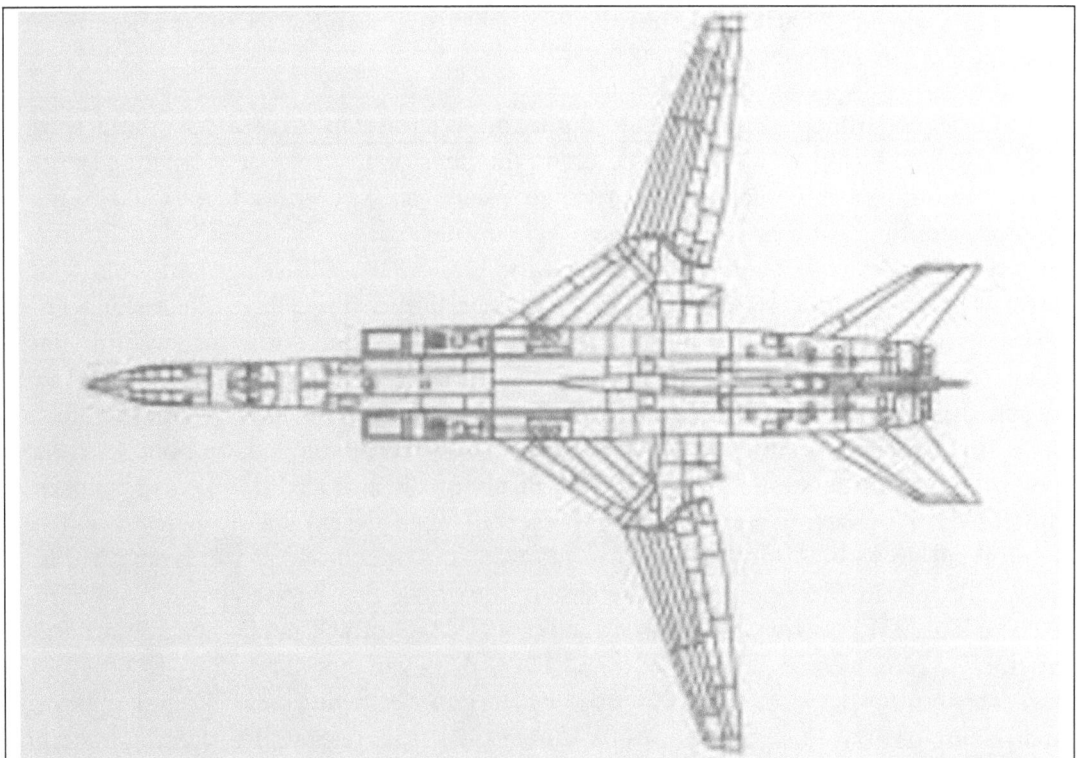

Previous page: Underside plan view of the Tu-22M from which the Tu-22M3 was developed. This page: Tu-22M3's operating with Russian Federation Long Range Aviation. Tupolev/UAC

Tu-22M3 of Long Range Aviation. UAC

Capable of carrying up to 24040 kg (53,000 lb.) of ordnance in the internal weapons bay and on external stores stations, the Tu-22M3 was tasked with a multitude of roles, including anti-ship strike, nuclear strike and conventional bombing (MODRF). The primary anti-ship weapon is the Kh-22 high supersonic ALCM, a single example of this large weapon being carried in a semi-conformal manner on the fuselage underside. A further two Kh-22 could be carried on stations mounted on the inboard, fixed wing section – these stations alternatively are capable of carrying other munitions. For the nuclear strike role the Tu-22M3 could carry the Kh-15 short range supersonic ALCM. In the conventional bombing role the Tu-22M3 is apparently capable of carrying up to 69 x FAB-250 series 250 kg class bombs or 42 x 500 kg weapons in the internal weapon bay and externally on multi-shackle racks, this being in excess of the number of 250 kg class weapons that could be carried by the considerably larger American B-52H strategic bomber, although the B-52H has a considerably longer range. Larger free fall weapons could be carried by the Tu-22M3, such as the FAB-1500 1500 kg class weapon, eight being the maximum load out. The FAB-250 series bombs appear to have been the standard weapon employed on operations over Syria, the normal load carried on missions appearing to be twelve such weapons. However, released video graphic material showed a Tu-22M3 dropping a single large weapon, presumably a FAB-1500 series bomb unit.

The prototype Tu-22M0 had been manufactured by mid-1969 and conducted its maiden flight on 30 August that year (test pilot V.P. Borisov). The first Tu-22M3 conducted its maiden flight on 20 June 1977 and, following testing, the design was ordered into serial production in 1978. Although having been in service for several years, the Tu-22M3 was officially accepted for service by Long Range Aviation of the Soviet Union in March 1989 (Tupolev).

Around 500 Tu-22M0/1/2/3 aircraft were built at the Kazan plant where series production had commenced in 1971. Exact number of Tu-22M3 built are hazy at best, but this is considered to be in excess of 200, with estimates of around 250 considered plausible (attempts to get an exact number from the manufacturer or designer to clear up the discretion failed to yield a definitive value), production ending in 1993. At the close of 1993 Long Range Aviation operated around 100 Tu-22M series with a further 185 operating with Russian Naval Aviation (MODRF). Recent western estimates put the figure of operational Tu-22M3/MR (the Tu-22MR is a reconnaissance variant of the Tu-22M3, only a small number being built, the survivors remaining in Long Range Aviation service in 2018) at around 70, although it is clear that considerably more airframes are available, some employed for training and trials work and others kept in various conditions of open storage. A number of Tu-22M series inherited by Ukraine were scrapped.

The Russian Federation Tu-22M3 fleet is classified as intermediate range bomber/missile carrier aircraft. However, the type can attain strategic range through the use of in-flight refueling from Long Range Aviation's fleet of Il-78M in-flight refueling tanker aircraft. This aircraft is shown returning to Kaluga, western Russia, from North Ossetia, where the aircraft had been based for bombing operations against Islamic State targets in the Syrian Arab Republic. MODRF

Tu-22M3 intermediate range bomber/missile carriers conducted large numbers of strikes on Islamic State targets in the Syrian Arab Republic from late 2015 through 2017. A Tu-22M3 is escorted by a Sukhoi Su-30SM multi-role strike fighter (top) and a pair of Tu-22M3's release bomb loads of 12 x FAB-250 250 kg unguided bombs against designated targets (above). MODRF

2

MODERISATION OF THE RUSSIAN STRATEGIC MISSILE CARRIER/BOMBER FLEET

In the second decade of the twenty first century the Russian Federation Long Range Aviation strategic missile carrier/bomber fleet has taken on a central role in that nations maintenance as a formidable military entity in the face of perceived threats from NATO (North Atlantic Treaty Organisation) encroachment (not violation) on her borders. The Tupolev Tu-95MS and Tu-160 fleets have, since 2007, reestablished routine long distance patrol flights into their training schedules. This practiced increased in intensity in the second decade of the twenty first century. For instance, in February 2011 a Tu-95MS conducted a 15 hour long range patrol in the area of the Norwegian Sea. Other Tu-95MS conducted long range patrols over the Arctic, North Atlantic and Pacific Oceans and the Black Sea (MODRF). Tu-160's have made regular patrols into the Norwegian and North Seas and, despite the fact that they fly in international airspace and abide by the normal rules governing such operations, major western media outlets go into hysterical overdrive when such patrols occur – even to the point of reporting overflights of the British Isles that did not actually take place. In line with the Arctic's increased importance to Russia's security and economic health a number of bases have been developed in Russia's Arctic North. In August 2018, a Tu-160 training mission was flown from Central Russia to Anadyr in the Russian Far East Arctic region where the aircraft landed. Ten aircraft were involved in this operation (it is unclear if this was the number of Tu-160's involved or if it included the Il-78M in-flight refueling tanker aircraft), which was the first time that Tu-160's had landed in the Arctic region. In December 2018, a pair of Tu-160's flew across the Atlantic to Venezuela, building on experience gained in such flights in 2008 and 2013.

By the time of the September 2015 Russian intervention against ISIS (Islamic State) and other extremist opposition groups attempting to overrun the Syrian Arab Republic, Long Range Aviation operated a fleet of around 28 Tu-95-6 and 35 Tu-95-16 strategic missile carriers, 16 Tu-160 strategic missile carriers and around 70 or so Tu-22M3 intermediate range missile carrier/bombers. The various fleets were undergoing overhaul and modernisations under Stage 1 overhaul and modernisation programs – a wider program to overhaul and update much of Long Range Aviation's Tu-95MS, Tu-160 and Tu-22M3 fleets. In regard to the Tu-95MS this included replacing the electronic equipment and improving the aircraft targeting capabilities, intended to keep the aircraft current until 2025 when, under 2016 planning, it was hoped the PAK DA (Perspective Aviation Complex for Long Range Aviation) would be available. Modernisation of the Tu-95MS Obzor-MS radar system, as is the case with the Obzor-K radar system installed in the Tu-160, was aimed at providing improved targeting and navigation capabilities and a reduction in detectability. For this latter reason the radar update was informally referred to as the low probability of intercept enhancement, although it is unclear to what extent the radar modernisation was implemented in the Stage 1 modernised Tu-95MS and Tu-160.

By mid-December 2015, eight or nine post-Stage 1 modernised Tu-95MS had been redelivered to Long Range Aviation. Several more were delivered in each of 2016 and 2017 years. During 2018, eight strategic missile carriers were expected to be redelivered post-Stage 1 modernisation, but the Russian defence Minister confirmed actual deliveries of 4 Tu-95MS and a single Tu-160 (MODRF).

In mid-August 2018, whilst the Stage 1 modernisation program was underway, it was confirmed that a further modernisation was planned to bring a portion of the Tu-95MS fleet to a standard referred to as Tu-95MSM. While at this juncture it was confirmed that the contract for the modernisation had been finalised, available details of the program were few and completely absent in regard the implementation schedule. There is evidence to suggest that the NK-12M(MV) power plant would be updated. To this end, a Tu-95MS equipped with a modernised power plant, centred on the PJSC Kuznetsov developed NK-12PM and driving SPE ASerosila developed AV-60T propellers, was flown on 19 April 2018. The new power plant was developed to increase range and payload and improve take-off characteristics. The aircraft was handed over to the Russian Defence Ministry State Flight Test Centre and embarked upon a series of state joint testing with the trials scheduled for conclusion toward the end of 2018 (Tupolev).

Page 30: Tu-95MS returned to Long Range Aviation on 29 December 2014 following Stage 1 overhaul and modernisation. This page top: Tu-95MS redelivered, post Stage 1 overhaul and modernisation, in 2015-2016. Above: Tu-95MS undergoing Stage 1 overhaul and modernisation. Tupolev/UAC

Tu-95MS redelivered to Long Range Aviation in 2014 following Stage 1 overhaul and modernisation. Tupolev

Top and above: A Tu-95MS, Red 25, was redelivered to Long Range Aviation on 27 February 2018 following Stage 1 overhaul and modernisation. Tupolev/Beriev

The Tu-160, referred to in Russian Federation Long Range Aviation service as an Intercontinental Strategic Multi-Mode Missile Carrier, is designed to strike high value installations on intercontinental deep penetration missions. The Tu-160 possesses enormous combat potential in both nuclear and conventional long-range strike scenarios armed primarily with air launched cruise missiles. Going into 2016, the Russian Federation operated a fleet of 16 Tu-160's, including a few Stage 1 modernised aircraft, the first of which had been redelivered from the Kazan Aircraft

Factory on 19 December 2014. The Tu-160 Stage 1 overhaul and modernisation program, which incorporated an updated avionics suite that had completed bench testing on 25 March 2013, was still ongoing into 2018. A Tu-160 was returned to Long Range Aviation on 28 August 2017 and another was redelivered post-Stage 1 modernisation in 2018 (Tupolev).

Top: A Tu-160 shadowed by a RAF Eurofighter Typhoon. Above: A Tu-160 redelivered to Long Range Aviation on 19 December 2014. UK MoD/Tupolev

In 2015, the year the Tu-160 was employed on operational missions over the Syrian Arab Republic, the MODRF (Ministry of Defence of the Russian Federation) authorised development of the Tu-160M2. The intention with the Tu-160M2 (previously referred to as the Tu-160M) program was to produce a completely new aircraft in terms of onboard systems. The airframe and engines would be modernised and the design would take into account modern design techniques. In 2014, Kuznetsoz Enterprise in Samara, Russia, had commenced development of a modernized NK-32 engine for the prospective Tu-160M2. The Russian Deputy Defence Minister confirmed that the modernised engines were ~10% more efficient compared with the legacy NK-32 of the Tu-160. This bestowed upon the Tu-160M2 a planned design range increase in the order of 1000 km over its forebear. It was also confirmed by the Deputy Defence Minister that the Tu-160M2 observability in the radio, and possibly the infrared spectrum's, would be reduced over that of the Tu-160 due to application of 'special coatings' designed to absorb radio waves, reducing the return to the illuminating radar (MODRF).

A new build Tu-160 is maneuvered from the final assembly hanger at Kazan in November 2017. Tupolev

The Tu-160M2 and the Tu-22M3M, detailed below, are equipped with a modern integrated fight control system with quadruple redundancy in the shape of the KSU-130 that was developed for the Yakovlev Yak-130 jet powered advanced trainer/light combat aircraft. The most significant changes in the Tu-160M2 over the Tu-160 is in the realm of the radio-electronic/sensor/avionics suites (Kret). It is known, through public statements and releases, that the Tu-160M2 will be equipped with an IMA

(Integrated Modular Avionics) suite. Enhancing the designs capabilities is a 'computational and on-board systems, control facilities, a fundamentally new navigation system, a highly effective electronic warfare complex, and weapons control systems' (Kret). The new equipment developed by Kret enterprises are designed not only for the Tu-160M2, but also for possible retrofit to the existing Tu-160 fleet and some elements have been incorporated into the Tu-22M3M modernisation program (Kret). A variation of the United Instrument Manufacturing S-111 communications complex equipping the Sukhoi Su-57 fifth generation fighter aircraft has been specified for the Tu-160M2, Tu-22M3M and PAK DA (Rostec).

In 2018, the commander of Russian Long Range Aviation, Sergey Kobylash, confirmed that the Tu-160M2 and the Tu-22M3M would have basically the same radio-electronic suite. Furthermore, Kret expects that some elements of the IMA developed for the Tu-160M2 will be incorporated into the design of the PAK DA under development simultaneously with the Tu-160M2 (Kret). The design improvements and modern radio-electronics/avionics are designed to improve combat effectiveness by around 2.5 times (also stated as the lower value of around 60%) in comparison to the Tu-160 (Kret).

A new build Tu-160 is maneuvered from the final assembly hanger at Kazan in November 2017. Tupolev

Summer 2013 planning, as confirmed by PJSC Tupolev General Director, Alexander Konyuhov, called for all operational first generation Tu-160's to be modernised to Tu-160M (later Tu-160M2) standard (Tupolev). The planned schedule at that time called for the first flight of the prototype Tu-160M in the 3rd or 4th quarter of 2019, the transfer of the first new build Tu-160M to the VKS (Russian Federation Air Force) in 2021. The first serial upgraded Tu-160M, from Tu-160 standard, was scheduled to be transferred to the VKS in 2022. With the finalisation of plans for manufacture of the updated aircraft in 2015, by which time it was designated Tu-160M2, several statements of revised schedules emerged. In a visit to the Kazan Aviation Factory on 8 June 2017, it was announced, by the Russian Deputy Defence Minister, that the first welding of components for the first Tu-160M2 had taken place (MODRF). At the same time the MODRF confirmed the intention to acquire 30-50 Tu-160M2 and that serial production was now scheduled to commence in 2022 (Kret). As of August 2017, it was intended to fly the first Tu-160M2 by the end of 2019, with series production then expected to commence sometime in 2021 (MODRF). The UAC (United Aircraft Corporation) head stated 'Experimental design work for us ends in 2021, starting in 2022, we begin mass production of these machines [Tu-160M2]' (Kret). In 2018, this schedule was little changed – 2019 for the Tu-160M2 prototype maiden flight and 2021-2022 for commencement of serial production.

In March 2018, it was stated by the Russian Deputy Defence Minister that Russia would completely renew the Tu-160 fleet by 2030. The minister went on to say 'we are going to purchase the entire fleet of... Tu-160 [M2] bombers... and carry out heavy upgrade of operational aircraft where only the fuselage will remain while all onboard radio-electronic equipment and engines will be replaced'. This statement appeared to confirm, at Russian government level, Tupolev's previous assertion that in addition to the new build Tu-160M2s, all, or most, of the current (2018) fleet of Long Range Aviation Tu-160's would be modernised to Tu-160M2 standard. What is unclear is whether or not the quoted value of 30-50 Tu-160M2's includes those Tu-160's to be modernised to Tu-160M2 standard or if this value refers only to procurement of new build aircraft.

A new build Tu-160 is maneuvered from the final assembly hanger in November 2017. Tupolev

The program to recommence production of the Tu-160 was initiated in 2005. The program stagnated, but was eventually given renewed impetus with the advancement of the Tu-160M/2 program in the second decade of the twenty first century. On 16 November 2017, a new production Tu-160 was rolled out following assembly at the Kazan Aviation Factory (Tupolev). The primary reasoning behind manufacture of this aircraft before the implementation of Tu-160M2 production was to re-familiarise the plant with Tu-160 manufacturing and assembly techniques in preparation for production of the Tu-160M2. New design/manufacturing processes were introduced, employing digital technologies. This new build Tu-160 conducted its maiden flight on 26 January 2018 (Tupolev).

In late summer 2018 the Tu-160M2 and Tu-95MSM programs were described as priority areas for the overall modernisation of the Russian armed forces. This stems from their not only being an integral part of the Russian nuclear deterrent triad, but also their ability to strike NATO targets at long-range in the event of conventional conflict in Europe or the Far East.

A new build Tu-160 conducted its maiden flight on 28 January 2018, commencing a short acceptance flight test phase. Tupolev

Commencing in the early part of the second decade of the twenty first century the operational Tu-22M3 fleet was put through the Stage 1 overhaul and modernisation program. A number of these aircraft were subsequently involved in the Russian air campaign over the Syrian Arab Republic with sortie numbers in the triple digits flown from South Ossetia – this mission ended in November 2017. The Stage 1 program was still ongoing in 2018, an overhauled Tu-22M3 being handed over to the MODRF on 3 April that year (Tupolev).

Previous page top: A Tu-22M3 undergoing Stage 1 overhaul and modernisation at the Kazan Aircraft Factory. Previous page bottom and this page: Tu-22M3's (Blue 14 (previous page, Red 16 this page top and Blue 42, above) being redelivered to Long Range Aviation following Stage 1 overhaul and modernisation at the Kazan Aircraft Factory in 2015. Tupolev

In the modernisation of the Russian Federation armed forces the Tu-22M3 has taken on a central role in Russia's ability to project power in defence of the homeland or in support of allied countries such as the CSTO (Collective Security Treaty Organisation) nations or the Syrian Arab Republic. As well as the combat operations flown over Syria, the Tu-22M3 bomber fleet has been employed in training operations that simulated strikes on extremist targets on the Tajik-Afghanistan border region, supporting the CSTO CRRF (Collective Rapid Reaction Force). One such operation, which took place in mid-November 2017, involved an undisclosed number of Tu-22M3's, which operated out of Tolmachyovo airbase, Novosibirsk. These aircraft released FAB-250 250 kg high explosive fragmentation bombs over notional terrorist group positions located in the Kharbmaydon Mountain Range in the Tajik-Afghanistan border region. The Tu-22M3's were escorted by RAC-MiG MiG-31 interceptors operating from Kant airbase in Kyrgyzstan, to which they had transferred from Krasnoyarsk Krai for the operation (MODRF). A similar operation had been conducted on 1 July 2017 when Tu-22M3's were transferred from Shaykovka to Tolmachyovo for participation in the exercise 'Dushanbe-Anti-Terror-2017' (MODRF). In this operation Tu-22M3's dropped FAB-250 bombs on a notional terrorist target in the Kharb-Maydon mountain range some 15 km from the Tajik-Afghan border (MODRF). This theatre is of importance to the CSTO nations due to the threat of Islamic extremists forming on the Afghan side of the border, combined with the NATO presence in Afghanistan and the inter-related unrest in the Tajik-Afghan border region.

A Tu-22M3, Blue 21, being redelivered to the MODRF post-Stage 1 overhaul and modernisation on 30 November 2017. UAC

In 2018 Long Range Aviation operated a mixed fleet of pre and post Stage 1 overhaul and modernisation standard Tu-22M3 intermediate missile carrier/bombers. MODRF

Top: Tu-22M3 participating in the Russian Federation AVIADARTS 2018 manoeuvres. Above: A Tu-22M3 operating from the Belaya air base, Irkutsk region of Russia, during the Vostok (East) 2018 large scale military exercise held in September 2018. MODRF

The success of the Tu-22M3 operations in the Syrian Arab Republic, where it was credited with destroying ISIS offensives in the Deir-ez-Zor region and ISIS controlled oil facilities, outlined the designs importance to expeditionary warfare scenarios and the defence of the Russian Federation from direct attack. The types importance in the over land strike role, combined with the strategically important role of defending the Russian Federation against maritime threats – aircraft carrier battle groups and surface cruise missile carriers – led to various studies for capability enhancements and modernisation that went considerably beyond that implemented in the Stage 1 overhaul and modernisation program. The studies would lead directly to the Tu-22M3M modification of the existing Tu-22M3. The Tu-22M3M upgrade had been designed by November 2017, allowing preparations for the commencement of modification of the first Tu-22M3 to M3M standard.

When the program was revealed it was stated that the Tu-22M3M and the Tu-160M2 would employ a number of unified systems (Kret). While few details of the Tu-160M2 modernisation have been revealed more details have emerged of the systems installed in the Tu-22M3M. However, while the temptation is omnipresent to apply this information back to the Tu-160M2, caution is advised as the two designs are may not have the same systems across the board. What we know, in 2018, is that at the heart of the Tu-22M3M modernisation is the incorporation of new 'digital on-board radio electronic equipment' (BREO) taken from available Russian developed technology (Tupolev). The majority of analogue systems (around 80%) available in the Tu-22M3 have been replaced by digital systems in the Tu-22M3M, bestowing upon the latter improved capabilities in the areas of accurate weapon delivery, navigation in complex environments and increased automation, reducing prefight preparation and routine maintenance requirements (Tupolev). As well as the radio-electronic complex, which is stated to be identical to that planned for the Tu-160M2, the modernisation also incorporates new advanced navigation and communications systems – confirmed by the head of PJSC Tupolev – and a modern weapon control system and data control system. Specific items specified for the Tu-22M3M include the SPV-24-22 digital specialised computer sub-system, which allows full use of the GLONASS (Global Navigation Satellite System), this complex is also thought to have been incorporated into the Tu-160M2 design. The ability to conduct precise bomb strikes when employing unguided munitions is enhanced through the incorporation of an ordnance aiming navigation system known as Gefast. Other areas of the modernisation include incorporation of a modern IMS (Information Management System) and modern digital display screens being incorporated into modernised four-crew cockpit sections (Tupolev). A new digital control system for the engines and fuel supply, combined with improvements to the NK-12 engines, allowed unrefueled range to be extended (no value provided) over that of the Tu-22M3 (Tupolev). At the time (2018) of confirming the that the same radio-electronic equipment suite would be incorporated in the Tu-22M3M and Tu 160M2, the commander of Russian Long Range Aviation, Sergey Kobylash, also confirmed that the Tu-22M3M would have AI (Artificial Intelligence) systems incorporated into the design.

The Tu-22M3M program incorporates a service life extension that extends to 45-50 years (Tupolev). For a Tu-22M3 aircraft built in 1991 this would extend service life out to 2036-2041.

Page 44-45: Roll out of the prototype conversion of the Tu-22M3M in summer 2018.
Tupolev

The Tu-22M3M armament options include guided and unguided free fall ordnance and guided missiles (Tupolev). Although the Tu-22M3M will retain an inland strike role, the primary mission will be the elimination of sea surface combat groups threatening the Russian homeland – aircraft carrier battle groups and surface groups armed with land attack cruise missiles. This sea surface denial mission is still the domain of those Tu-22M3's administered under Russian Naval Aviation. The primary armament planned for carriage by the Tu-22M3M will be the X-32 (Kh-32) high supersonic cruise missile (this is apparently a deep redesign of the high supersonic Kh-22M series), which is expected to be carried in a similar fashion to the X-22 (Kh-22) carriage on the Tu-22M3. In addition, the head of the Federation Council Defence Committee and former commander in chief of the Russian Aerospace Force, Viktor Bondarev, confirmed plans to arm the Tu-22M3M with hypersonic missiles – taken to mean the Kh-47 Kinzhal then in limited trials service with MiG-31K's of the Russian Federation Southern Military District. In 2018, the Tu-22M3M/Kh-47 is apparently the subject of a feasibility study to ascertain the formers suitability as a launch platform for the latter.

The Tu-22M3M prototype, a modified Tu-22M3, was rolled out at Tupolev's Kazan Aircraft Factory on 16 August 2018, following which the aircraft was taken in hand for a series of ground testing prior to being handed over to the Ministry of Defence of the Russian Federation for participation in State Joint Tests. In October 2018 the maiden flight was delayed. At the same time it was revealed that the ground test phase of the AI systems was still ongoing, suggesting that this was the area, or one area, of development/integration responsible for the delay to flight testing (unconfirmed in 2018). The maiden flight was conducted from a snow covered Kazan on 28 December 2018. The final decision on whether or not to proceed with the modernisation of 30 Tu-22M3 to Tu-22M3M standard is to be taken based on the results of the trials of the prototype (Tupolev).

Externally the Tu-22M3M is identical to the Tu-22M3 with a few exceptions, notably the inclusion of a long fairing on the upper nose section (top) and removal of the GSh-23 cannon armament complex from the tail (above) of the former. Tupolev

Tu-22M3M roll out in summer 2018 (top) and lift off from Kazan on its maiden flight on 28 December 2018 (above). UAC

A Tu-95MS-16 carrying a single Kh-101/102 on the inner launch station of the port and starboard outer missile carrier beams. MODRF

The Tu-95MS and the Tu-160 began equipping Long Range Aviation in the last decade of the Cold War. Tasked with the long-range nuclear strike role, the major armament for both strategic missile carrier types was the X-55 (Kh-55) long-range nuclear armed air to surface cruise missile. The X-55 had been developed by ICB Rainbow/GosMKB Raduga (designers on the project included I.S. Seleznev (General Designer), V.A. Kovalchuk, V.A. Pavlov, R.A. Tartsev, L.V. Bogolyubsky, R.Sh. Khaykin, K.N. Subbotin, M.M. Galperin, V.N. Kucheryavey, V.N. Truson, G.A. Vershinin, L.A. Mavlyanov, A.I. Dmitriev, N.B. Makarov & V Lyapunov) by 1976 and was intended for carriage not only by the then planned fleet of Tu-95MS and Tu-160, but, in the interim and for trials, by the Tu-95M-55 (DBM). The first launch of an X-55 from a Tu-160 took place in June 1987.

The basic X-55 airframe is described as a glider of 'all-metal, welded monoplane construction' (DBM). The engine extends downward from the rear fuselage through an opening once in flight and small wings are folded into the fuselage sides (at the 'tank compartment') (DBM) when stored or prior to launch and are extended for cruise flight. Other control surfaces include the retractable unit. The central tank compartment, which has an access area under the wing section, incorporates the housing for the special (nuclear warhead) payload (DBM). The forward monolithic fairing was constructed of 'hollow silicon-organic cloth' – 'K-9-70 binder' (DBM). The guidance system incorporates an INS (Inertial Navigation System) combined with a terrain correction (avoidance) system for flight at very low altitudes (DBM).

Изделие «120»

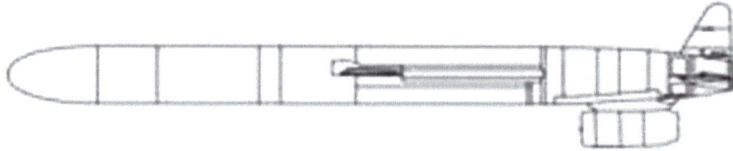

Дозвуковая крылатая ракета

Предназначена для поражения стационарных объектов с заранее известными координатами.

Russian language graphic depicting the X-55, which was initially known by the factory code Product 120. Text reproduced below with English translation in parenthesis: изделие 120 (Product 120); дозвуковая крылатая ракета (subsonic cruise missile); предназначена для поражения стационарных объектов с заранее известными координатами (designed to destroy stationary objects with previously known coordinates). DBM

There were reports of a conventional armed variant of the X-55 designated X-555 (the designation Kh-550 has also been used at Russian Federation government level), which this researcher was admittedly skeptical existed in more than experimental form. However, despite reference only to the Kh-101 long-range conventional warhead air launched cruise missile being employed by the Tu-160 and Tu-95MS over Syria, there is clear graphic evidence of another missile, derived from the X-55, being launched from Tu-95MS. The existence of this weapon has since been confirmed at senior level in the Russian Federation government as the Kh-550 although the official line remains that the Kh-101 was the only air launched cruise missile employed by Russia during the campaign in Syria.

Graphic data shows a weapon completely distinct from the Kh-101 being launched from the Tu-95MS. This missile could be the conventional derivative of the Kh-55, often referred to as the Kh-555 (550) or perhaps a completely distinct weapon. MODRF

In the second decade of the twenty first century new generation subsonic air launched cruise missiles entered the inventory of Long Range Aviation. For the primary nuclear strike mission the nuclear armed Kh-102 has supplemented the Kh-55. In the conventional long-range strike role the Kh-102 is replaced by the conventional armed Kh-101, which is apparently more or less identical to the Kh-102 with the exception that the nuclear warhead in the latter is replaced by a 500 kg class high explosive warhead in the former. As noted above the Kh-101 is the only weapon acknowledged by the MODRF as being employed by the Tu-95MS and Tu-160 on operations over Syria. Very little is known about the Kh-101/102, which were designed as a low-observable air launched cruise missiles. These weapons are often simply described as modernised variants of the Kh-55 air launched cruise missile. On analysis, however, the Kh-101, which apparently entered Long Range

Aviation service in late 2012, appears to be much more than simply an evolved Kh-55, range being more than doubled from the 2000 km of the Kh-55 to ~4500 km (MODRF). In addition, the contoured flattened body shaping is distinct from the cylindrical design of the Kh-55. The Kh-55 was designed to be difficult to detect in the radio contrast spectrum, the low observable qualities of the Kh-101/102 being considerably enhanced, making the missile very difficult to detect and target in the visual, radio contrast and infrared spectrums. Like the Kh-55, the Kh-101/102 are powered by a single turbofan engine located in the rear of the missile body; this being extended below the missile just before launch. The small wings and tail surfaces do not deploy until after the missile has been released from the rotary launcher and drops away from the carrier aircraft.

Kh-101 air launched cruise missiles on a rotary launcher in a weapons bay on a Tu-160. MODRF

Kh-101/102 air launched cruise missiles being loaded onto a Tu-160 in 2018. MODRF

In regard to the Kh-55, the missile follows a terrain following profile between 30 and 100 m from the Earth's surface, cruising at speeds of ~850 km/h. For over water operations an INS is employed. No details have been released on the guidance systems employed by the Kh-101/102, but these certainly include a global positioning system combined with a modern terrain following system. In the Kh-55, as is likely the case with the Kh-101/102, targeting and terrain profile data was uploaded to the missile before it was loaded onto the launch aircraft. Once launched and on their way to the target the missiles, launched in multiples of up to six, could change position regularly and profile mask, by which one missile maneuvers behind the other to confuse detection radar and electro-optic systems. On nearing the target the missiles can spread out further and can attack from different directions.

In 2017, news of a new generation cruise missile, intended for carriage by the Tu-160M2 and possibly the PAK DA, emerged. This weapon, referred to as the XB-D, has been described as a long-range stand-off cruise missile. Although no target range value has been released, the XB-D projected range is stated as being in excess of the 4500 km range of the Kh-101/102, air launched cruise missile (Kret). The deputy Defence Minister of the Russian Federation confirmed that new, more capable, air launched missiles were being developed for the Tu-160M2, stating '… you can't compare the Tu-160 plane with the Kh-55, Kh-550[555] and even the Kh-101 missiles with the plane [Tu-160M2]' armed with the prospective XB-D. The few details of this new missile program fail to address whether or not it would be a subsonic missile like the Kh-55/550(555)/101/102 or whether it would be capable of supersonic or even flight hypersonic speeds.

The high supersonic flight speed X-15 air launched cruise missile, developed by ICB Rainbow/GosMKB Raduga (designers on the project include I.S. Seleznev (general designer), B.V. Kulikov, V.A. Larionov, A.P. Chernikov, V.S. Shcheglov and S.Sh. Livshiys) by 1978, was developed for service with the Tu-22M3 (DBM). The X-15 was designed as a wingless glider design powered by a solid-fuel two-chamber rocket engine. The missile flew to the target area under the guidance of an internal control system, the missile body having few control surfaces – 'controlled cantilever[s]' (DBM). The instrumentation and payload compartment were constructed of OT4-1 and VT-5 material. Heat protection for the fuselage consisted of 'TZMKT and internal thermal insulation' (DBM).

The status of the X-15 within Long Range Aviation in 2018 in unclear, although, in the absence of evidence to the contrary, it is considered that the weapon still constitutes an element of the Tu-22M3 nuclear strike capability. However, it is unclear if this will be carried over to the modernised Tu-22M3M.

Kh-22/Kh-32 – The Tu-22M series would carry X-22M (Kh-22M) series air launched cruise missile. The high supersonic capable Kh-22 bestowed upon the operator a potent anti-ship strike capability to counter NATO aircraft carrier battle groups threatening the Soviet Union – post-1991, the Russian Federation. The X-22M can also be employed against radiocontrast land targets. In terms of attack velocity it would be difficult for surface ships to defend against the Kh-22 even in the second decade of the twenty first century. The Tu-22M3M is slated to be armed with the Mach 4-5 capable Kh-32, thought to be a deep modernisation of the Kh-22M2. The Kh-32 is expected to completely replace the Kh-22M2, although no out of service date has been announced for the latter.

Tu-22M3 intermediate range missile carrier/bomber armed with a single Kh-22 on the starboard fixed wing section station. MODRF

Изделие «Д2»

Сверхзвуковая высокоточная крылатая ракета

Ракета для вооружения самолетов дальней авиации. Модификации ракеты предназначены для нанесения ударов по наземным и морским радиоконтрастным и площадным целям.

Russian language graphic depicting the Product D2 (X-22). Text reproduced below with English translation in parenthesis: изделие Д2 (Product D2); сверхзвуковая высокоточная крылатая ракета (supersonic high precision cruise missile); ракета для вооружения самолетов дальней авиации (missile for armament of long range aircraft); модификции ракеты предназначены для нанесения ударов по наземным и морским радиоконтрастным и площадным целям (missile modifications are designed to attack sea and land radio-contrast and area targets. DBM

From its introduction to Soviet Naval Aviation in the mid-1970's the Tu-22M series took on a primary role of sea area denial armed with Kh-22 long-range anti-ship cruise missiles designed for attacks on high value assets such as **NATO** attack aircraft carriers. In 2018, the Russian Federation operates a force of Tu-22M3's in the missile carrier/bomber roles and expected to commence trials with the first updated **Tu-22M3M in late 2018/early 2019. This latter variant will apparently be armed with a new Mach 4-5 capable air launched cruise missile designated Kh-32.** US DoD/UAC

The Kinzhal Air System consists of an RAC-MiG MiG-31K launch platform and a Mach 10 (MODRF value) capable aero ballistic missile (Kh-47M2 – designation unconfirmed by MODRF as of late 2018) that is similar in appearance to the aero ballistic missile of the Iskander-M short-range ground-launched missile complex. The Kinzhal Air System entered limited service (10 launch platforms) with the Russian Federation Aerospace Forces in late 2017. MODRF

If the Kinzhal hypersonic cruise missile is introduced to widespread operational service then this will provide the Russian Federation with a counter to the NATO aircraft carrier battle group and intermediate range land attack cruise missile carriers, for which NATO would have no hard kill defensive capability against (acknowledged in a 2018 American threat assessment document). Current and projected attainable missile defense systems have no capability to counter hypersonic missiles attaining speeds of up to Mach 10. Furthermore, weapons in the class of Kinzhal and the lower velocity Kh-32 are designed with a maneuvering capability to further complicate defensive measures. Such a capability would defeat current (2019) NATO missile defense systems that are optimised to hit a lower velocity target following a predetermined ballistic trajectory. As noted above, in 2018, a technical study commenced to assess the feasibility of Kinzhal being carried by and launched from the Tu-22M3M. Assuming its viability it is expected that Kinzhal will constitute a major armament option on the Tu-22M3M fleet alongside the Kh-32.

In the Syrian campaign (2015-2017) the Russian missile carrier/bomber triad conducted their missions successfully. The Tu-160 and Tu-95MS were employed exclusively as missile carriers, but the Tu-22M3 conducted a largely unguided bombing mission. In the unguided bomber role the Tu-22M3 conducted its missions from medium altitudes with a high degree of accuracy. Such missions were fundamental in destroying ISIS oil production facilities and in providing high levels of ordnance against ISIS ground offensive in the relief of Deir ez Zor. Considering its importance in such operations it seems clear that unguided munitions will arm the Tu-22M3/M3M fleet beyond the second decade of the twenty first century.

Top: Top: Tu-160, RF-94113 (Red 17), launches a cruise missile against a target in Syria in 2015. In November 2018 a Tu-160 salvo launched a full complement of 12 Kh-101 missiles in salvo. Above: A night scene of the deployment of two Long Range Aviation Tu-160's to Venezuela in December 2018. This mission built on experience gained in two previous deployments in 2008 and 2013. MODRF

Top: Long Range Aviation Tu-22M3 missile carrier/bombers. Above: FAB-250 series bombs deployed from a Tu-22M3 operating from Mozdok air base, North-Ossetia, Russia, against ISIS targets in the Syrian Arab Republic. MODRF

3

PAK DA – PERSPECTIVE AVIAITON COMPLEX FOR LONG-RANGE AVIATION

With a maiden flight scheduled for some time in the first half of the 2020's, the PAK DA (Perspective Aviation Complex for Long-Range Aviation) is a new generation strategic missile carrier/bomber purportedly to replace the Tupolev Tu-95MS/MSM and Tupolev Tu-160/M2 strategic missile carrier fleets. Development of a new generation Russian (Soviet) strategic bomber can be traced back to the late Cold War period. In 1987, a CIA (Central Intelligence Agency) document outlined several descriptions, based on espionage reports, of a new Soviet bomber design. One source (079) indicated that the design lacked such equipment as an '(electronic) jamming system' or an advanced 'fuel filtering system' similar to that developed for the American B-2 strategic bomber (CIA, 1987). The source went on to detail 'a picture taking ability', indicating a reconnaissance role, and that the aircraft was, in 1987, in flight test, stating that it had not engaged 'in ... many flights' (CIA, 1987).

Another source (011) indicated that the Soviet bomber design was being tested in model form, but made no mention of a full-scale flying aircraft. The intelligence report stated that this model featured '...a flat and wide body with a leading edge that starts from the nose of the aircraft. It is propelled by two engines and makes use of liquid fuel gas under pressure that allow it to function out of the atmosphere' (CIA, 1987). This aircraft/spacecraft design, it was contended, was being developed under the auspices of the various Soviet Space organisations. The flattened wide body appearance of the scaled-model described appeared similar in description to another source (003) description of a full size aircraft with a 'flat squashed, appearance' (CIA, 1987). This source also indicated a design that sloped downward from the nose section toward the tail (CIA, 1987).

An analysis of the various sources of information regarding the Soviet bomber, dated 21 August 1987, held the following conclusion: 'The concept of this aircraft is radical and advanced. It is high tech and stealth-like. Its purpose is to operate within and outside of the e[E]arths atmosphere, capable of dual roles', the design being described as 'a cross between an SR-71 and a Space Shuttle)' (CIA, 1987).

Nuclear spacecraft M-19
1974

The existence of conceptual designs such as the Myasishchev Design Bureau M-19, which dated back to 1974, was instrumental in western intelligence agencies buying into bogus data coming from sources, apparently within the Soviet Union, that a new Soviet bomber, capable of atmospheric and exo-atmospheric flight, was being developed in the 1980's. EMZ

'I feel that the design and testing of this vehicle is not being conducted by the people or agency… responsible for the design of experimental aircraft. It is being designed and tested within the Space Program and at Space Research Facilities. 'For first signs of his vehicle keep an eye on the Space Research and Testing facilities.' (PSA, 1987).

Such an aero-spacecraft design did not show up on satellite imagery as it did not exist. The Soviet Union was, however, developing a reusable space transportation system under the Energia-Buran program. The resultant reusable spacecraft, named Buran, conducted a single launch into Earth orbit and returned to a conventional landing on Earth on 15 November 1988.

dual engines?

TOP VIEW

COCKPIT - (3-5 crewmembers)

air intake

Approved For Release 2000/08/08 : CIA-RDP96-00788R003800350003-6

INTERVIEWER NOTES:
1. SINGLE SAIL ASSEMBLY
2. STABILIZING GROOVES/RIDGES
3. GENERAL HOUR GLASS SHAPE

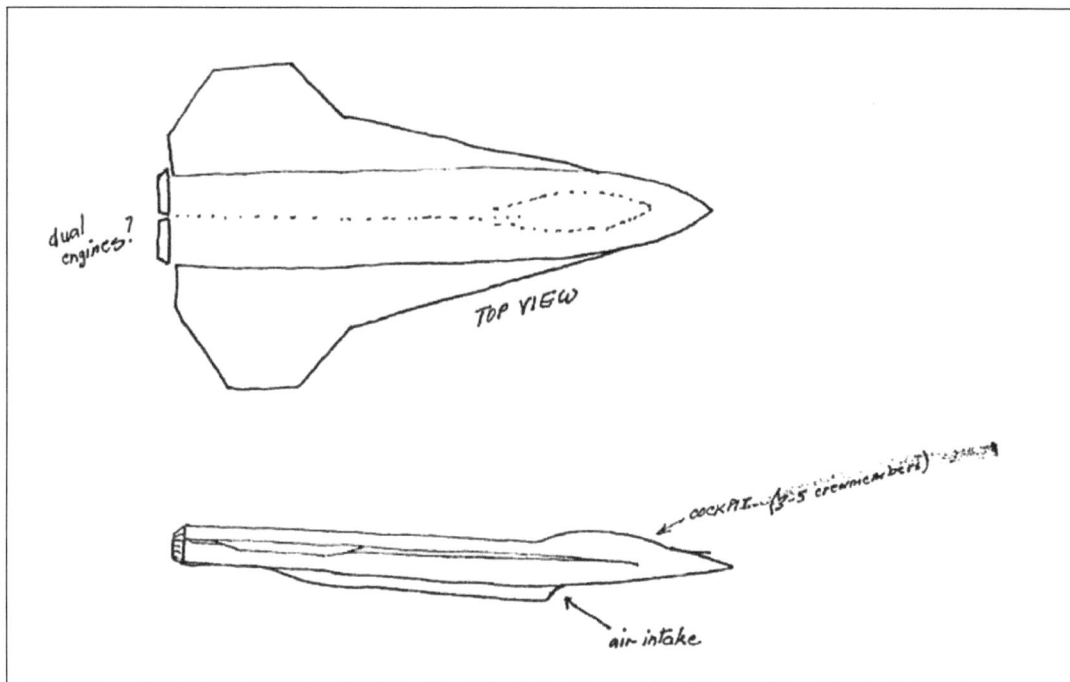

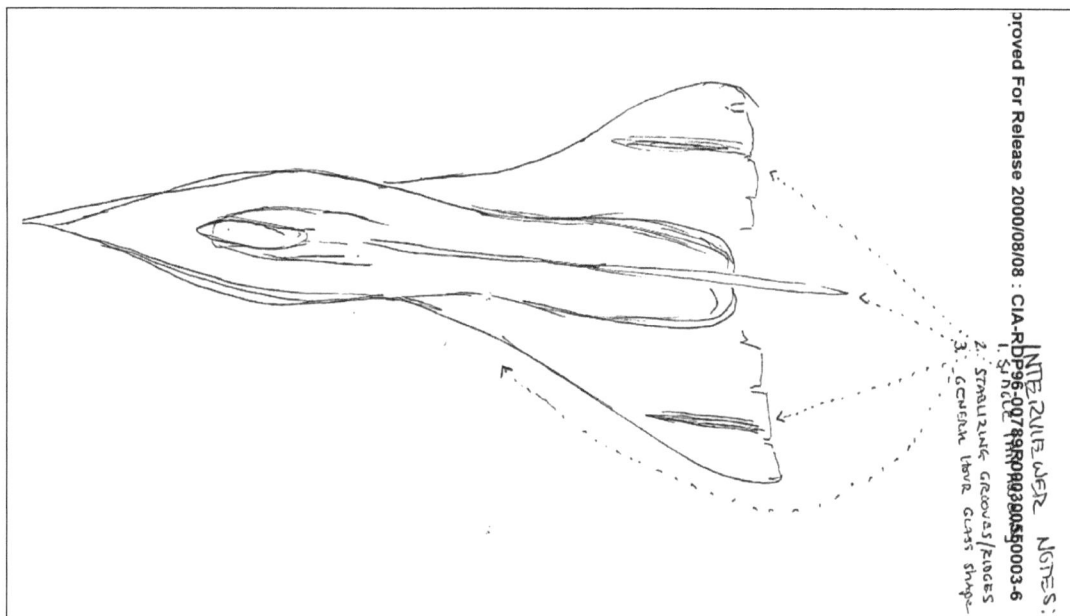

Top: Crude drawing, dated 21 August 1987, of a CIA assessment of a notional Soviet bomber design then thought to be under development. Characteristics include air fed to engines through under fuselage air intakes and a cockpit section to accommodate a crew estimated at 3-5 in number. Above: Another assessment resulted in this crude drawing with characteristics: 1. 'Single sail assembly'; 2. 'Stabilizing grooves/ridges;' 3. 'General hour glass shape'. CIA

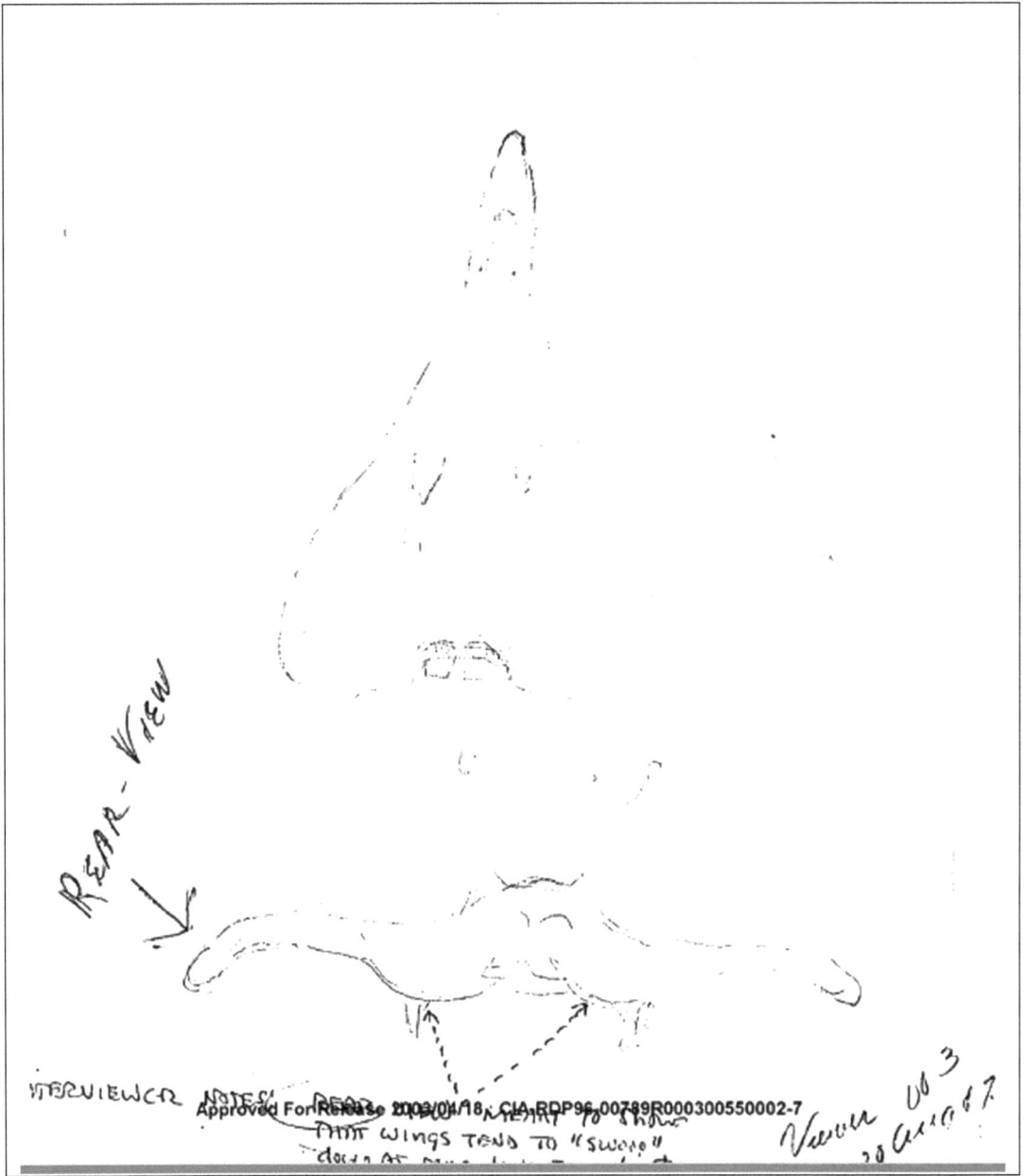

REAR-VIEW

Approved For Release 2003/04/18 : CIA-RDP96-00789R000300550002-7

INTERVIEWER NOTES: ... wings tend to "swoop"
down at ...

Crude drawing, dated 21 August 1987, of a CIA assessment of a notional Soviet bomber design then thought to be under development. Characteristics include: '…wings tend to "swoop" downward…' CIA

бомбардировщик-ракетоносец

ПАК ДА

2022-2025 г

Top: A Buran spacecraft conventional landing test article powered by turbojet engines. Such vehicles may have inspired the spurious claim among western intelligence agencies of an atmospheric/exo-atmospheric bomber. **Above:** Depiction of the бомбардировщик-ракетоносец (missile bomber); ПАК ДА (PAK DA) dated 2014 with flight test and serial production then slated for 2022-2025. Energia/UAC

Top: Flying wing configuration for a large four engine civil design tested at TsAGI during 2010. Above: Flying wing concept tested in the T-106 wind-tunnel at TsAGI in 2013. Testing was conducted at speeds up to Mach 0.88. TsAGI

An addition to the 21 August 1987 intelligence assessment, dated 25 August 1987, added that the Soviet bomber design featured a single vertical tail that was slopped rearward in a 'curved and recurved manner' (CIA, 1987). This was likened to an 'hourglass with two engines [located in the centre of the hourglass shape] (CIA, 1987). The August 1987 assessments indicated that the suspected Soviet bomber design, known in western intelligence agencies under the code name Project 8709, was an analogue of the American Northrop (now Boeing) B-2 'Stealth' bomber. Furthermore, the assessments, in areas, asserted that the design had been built and was in flight test. We know now that no such Soviet bomber program had been brought to fruition in the 1980's. More than three decades later an advanced bomber program is being prepared for flight in the Russian Federation – the major successor state of the former Soviet Union. Whilst there were, of course, a number of design studies for a new Soviet strategic missile carrier/bomber conducted in the 1980's and into the 1990's, such work all but ended following the dissolution of the Soviet Union on 25 December 1991. The embers of the bomber program were finally revived towards the second decade of the twenty first century in the form of the PAK DA program now underway.

In 2016/17, UAC (United Aircraft Corporation) confirmed that the PAK DA was being designed as a стратегический бомбардировщик-ракетоносец (Strategic Bomber-Missile Carrier) (UAC). Detailed design work on the PAK DA had commenced at PJSC Tupolev in 2009 (Kret). Many other enterprises, including Kret, commenced preliminary work on systems intended for the new bomber program. Kret provided Tupolev with various proposals for development of the on-board radio-electronic equipment suit intended for PAK DA (Kret).

Going into the second decade of the twenty first century the favoured layout for a future Russian strategic missile carrier/bomber was a flying wing concept. TsAGI was conducting research into new flying wing concepts for the future bomber design as well as for employment in the design of advanced new generation civil passenger aircraft. Through studies of a number of design concepts the most favoured approach was the flying-wing outline. Stage 1 testing of a flying wing concept was completed by the Aircraft and Missile Aerodynamics Department in late May 2013. This stage of testing involved tunnel tests, in the TsAGI T-106 transonic tunnel, of a model of a flying wing concept at speeds up to Mach 0.88 and large Reynolds-numbers corresponding to Mach 0.2. The test stage allowed for the aerodynamics of the concept to be refined. A number of options were tested with various engine and tail configurations that would be applicable to a range of aviation concepts ranging from civil airliners to strategic missile carriers. TsAGI had refined a flying-wing concept by July 2010. The model used in the 2013 T-106 tests was manufactured by TsAGI in 2011 and tested, with various engine and tail configurations, in the T-102 and T-107 subsonic tunnels at TsAGI in 2012 (TsAGI). While it is clear that these flying wing studies were conducted primarily in support of a future civil airliner concept, the evolution of the PAK DA over the period covering the first half of the second decade of the twenty first century may well be due to efficiencies in flight characteristics suggested by the 'flying wing' wind tunnel tests (TsAGI).

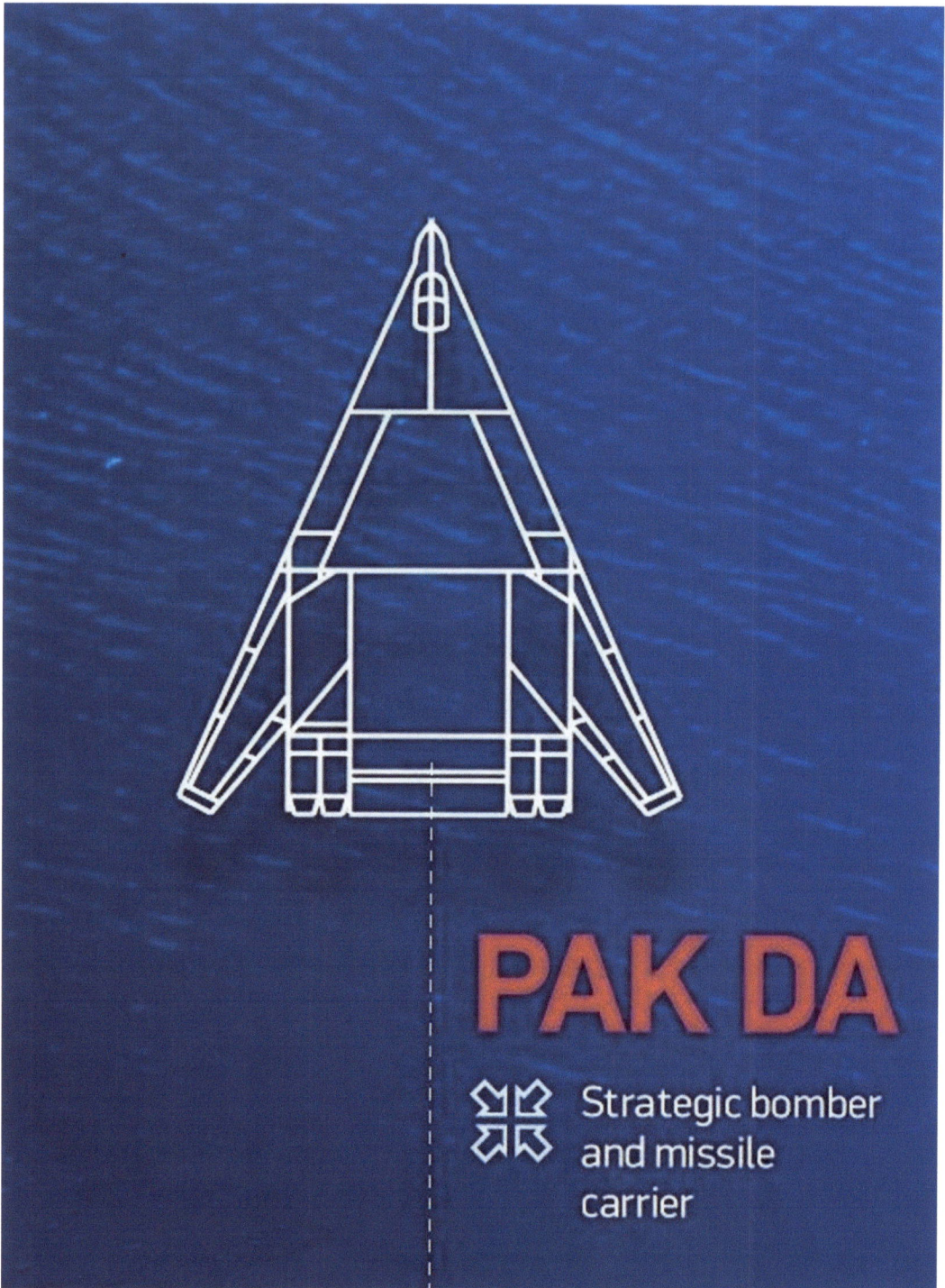

This c2015 graphic shows a PAK DA concept with four engines in two groups of two widely spaced apart on the under fuselage section. The concept could be interpreted to feature variable-geometry wings, although this may simply be illusory. Although flying wing concepts held sway for several years it has to be considered that this concept may have been purely notional. UAC

68

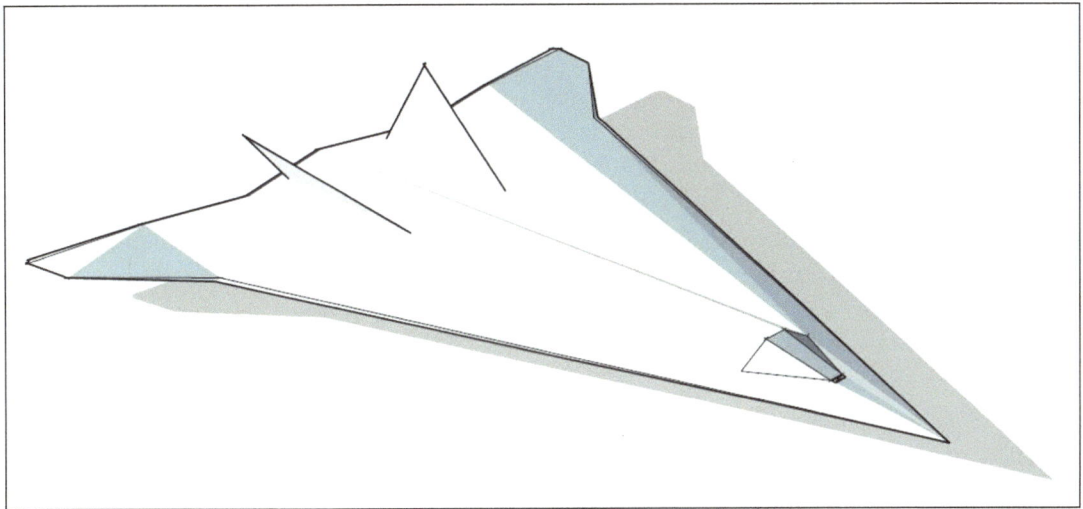

Two PAK DA concepts dating from 2014. The graphic (top) shows what is most likely a notional flying wing PAK DA concept with outward canted vertical tails. The graphic (above) outlines a design not dissimilar to the American B-2 'Spirit' strategic bomber design. UAC/Kret

The design studies for a 'flying wing' civil airliner concept had commenced in 1988, back in the Soviet era, around the same time that serious studies were initiated for a new strategic missile carrier/bomber for service with Soviet Long Range Aviation. Initially the Soviet flying wing civil airliner program was aimed at fielding a large high passenger capacity (up to 1,000 seats) aircraft. Into the twenty first century the flying wing concept was refined a number of times through the first decade until it arrived at a new target capacity of 150-200 seats. A similar evolution affected the future strategic missile carrier/bomber requirement. This was initially aimed at fielding a very large aircraft akin to something in the weight class of the Tu-95MS (this was still considerably lower in weight range to the Tu-160), but now appears to be aimed at fielding an aircraft more akin to the Tu-22M3 in regards to overall mass.

Although it had been conjectured for some time, that the PAK DA complex would be developed with stealth as a major design driver was officially confirmed by then Russian Federation Deputy Minister of Defence Alexey Krivoruchko on 13 July 2013. The aircraft was to be developed with low observability in the radio and infrared spectrums (Tupolev). This occasion was also taken to further confirm that Tupolev was the airframe prime developer (MODRF & UAC).

By 2014, work on the first stage of the 'detailed technical design' was being conducted according to the customer, MODRF (Ministry of Defence of the Russian Federation), approved schedule (UAC). The design phase of the PAK DA program was described as being in the initial stages in 2016-2017, cementing the move to a smaller mass aircraft. Considering that the Tu-160M2 heavy missile carrier program is being implemented in a similar timeframe, the PAK DA program may more appropriately be considered a Tu-22M3 replacement program.

The contract between UEC (United Engines Corporation) and PJSC Tupolev for development of the engines for the PAK DA was finalised in June 2018. At this time some basic operational details began to emerge. The engines, designed for routine operations for periods in excess of 30 hours, were to be capable of routine operation in a temperature range of -60° to +50° C and would be shielded from the effects of a nuclear detonation, although this ambiguous statement leaves open the question of whether this refers to EMP (Electro Magnetic Pulse) only.

The PAK DA was to be designed for operations from several airbase types, including the ability to use short runways and austere air bases, this being confirmed in August 2017 by then Russian Deputy Defence Minister, Yuri Borisov (Kret). This would reduce the PAK DA vulnerability to a first strike, assuming dispersal was carried out in times of increased threat.

We have, in in 2018, few details of the advanced sensor systems that will be incorporated into the PAK DA. In 2017, it had been confirmed that much of the avionics suite planned for installation in the Tu-160M2 would be carried over to the PAK DA, vastly reducing the costs associated with the PAK DA development (if such systems that are installed on the Tu-160M2 were in a actuality developed for the PAK DA then it is the Tu-160M2 that benefited from development of the PAK DA development effort) (MODRF). The ambiguous statements to the effect that some of the advanced systems to be incorporated in the Tu-160M2 will be carried over to the PAK DA is of little value until more is known about the sensor suite intended for the Tu-160M2. In regards to the primary radio-electronic sensor this is expected to be in a variation of the complex developed for the Tu-160M2. However, in 2017 it was speculated by Kret that a new targeting & navigation complex would be developed for PAK DA, as would communications, reconnaissance and electronic warfare system (Kret). The radio-electronic complex may well build on advanced technologies developed for Russia's 4++ and 5th generation tactical combat aircraft – Su-34, Su-35S and Su-57 (PAK FA – Perspective Aviation Complex for Frontline Aviation). Research programs will feed into the new radar complex, possibly including the BLRS polarisation radar research program revealed by Kret in June

2017 – this research program, conducted by the State Ryzan Instrument Plant, adopts the principles of polarisation selection to better determine stationary targets against the background of the Earth's surface' (Kret). It is clear that the PAK DA sensor suite will be optimised to allow strikes with advanced precision guided munitions against targets at long-range in day/night fair and adverse weather conditions.

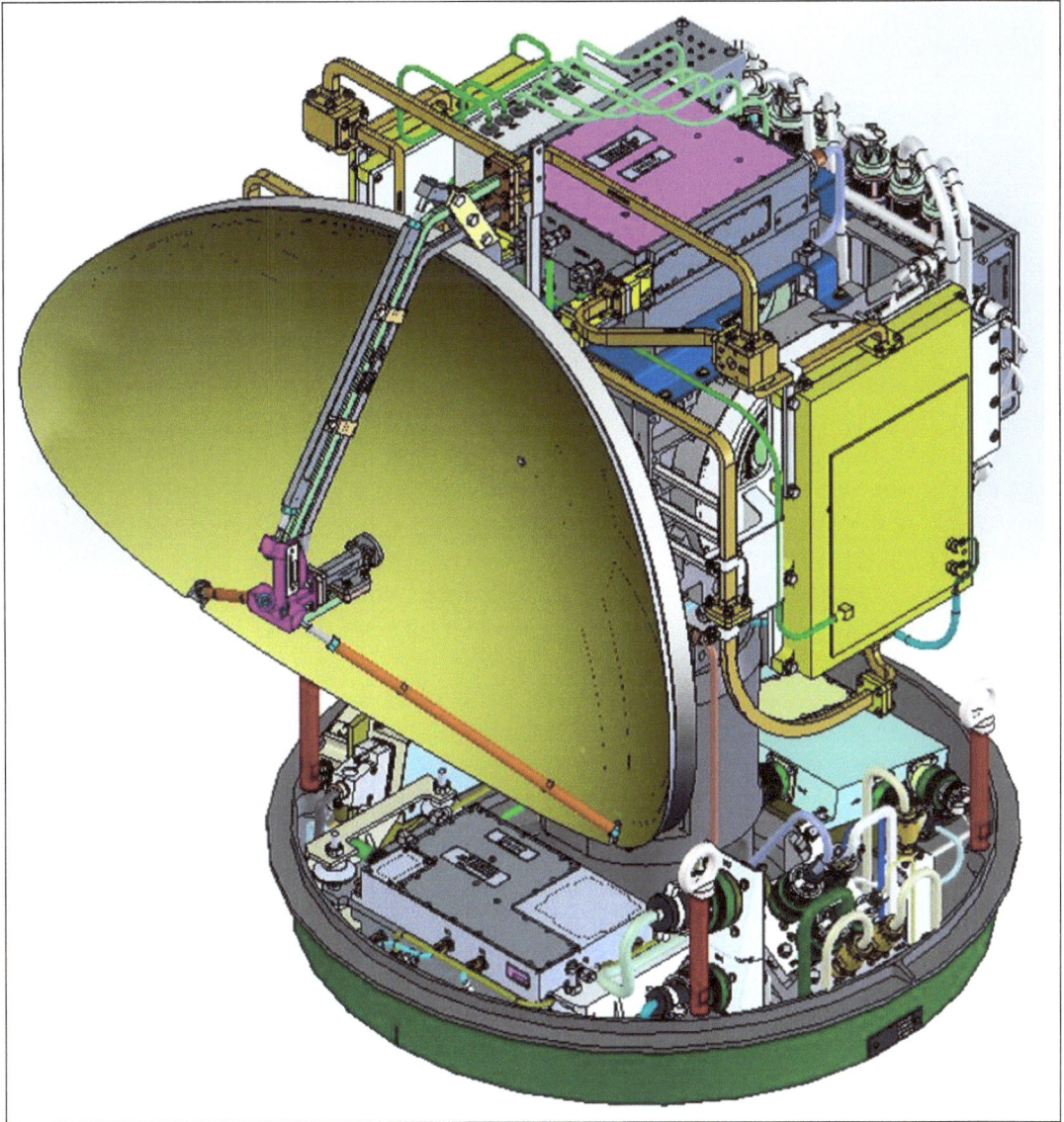

An airborne radar research program that may possibly feed into the PAK DA radio-electronic suite is the BLRS polarisation radar research program revealed in June 2017. Kret

The PAK DA radio-electronic suite remains something of an enigma in 2018, but it may possibly incorporate technology developed for the new generation of airborne tactical fighter aircraft radars typified by the Irbis-E and AFAR (APAR - Active Phased Array Radar) or be an unmodified variant of the sensor developed for the Tu-160M2. Tikhomirov

The intention with PAK DA is to arrive at a complex that can better overcome modern air defence systems when operating in adverse weather/climatic conditions than can current generation aircraft (Kret). The complex is intended for overland strike (nuclear and conventional) and maritime strike against sea surface targets (Kret), ports and naval bases. In 2018, to state the weapon complexes to be integrated with the PAK DA would be pure conjecture. However, it would be prudent to assume the possibility, if not probability, that some of the current generation weapons – X-32, X-101/102 – would be carried over, along with whatever complex emerges from the XB-D long-range new generation missile. What is known is that, in the design iteration extant in 2017-2018, the weapons would be carried in one or more internal weapons bays (MODRF). The operational load was described as good, leaving it open to speculation what value this would correspond to.

The importance of the weapons capability over aircraft performance in regard to flight speed was hinted at, indicating that subsonic speeds remained at the fore of the PAK DA design. MODRF information released in 2017 was the first to put forward the probability that the PAK DA would be armed with hypersonic missiles of very long range and that the enhanced ability to strike targets at long range (this could in many scenarios be from within the Russian Federation protected borders, further reducing vulnerability as the aircraft could avoid enemy air defence zones) was taking precedence over certain performance traits of the carrier aircraft itself (MODRF).

A Kinzhal hypersonic cruise missile is carried aloft (top) and launched from an RAC-MiG MiG-31K (above). MODRF

In 2013, it was confirmed that the PAK DA would be built at the Kazan Aviation Factory following an assessment of workload in regards to maintenance and modernisation of Tu-95MS, Tu-22M3 and Tu-160 strategic missile carrier/bombers and the planned manufacture of new production Tu-160M(2) strategic missile carriers (Tupolev). In July 2017, Kret announced that the intention was to fly the prototype of the PAK DA complex in 2025-2026 (this was revised from the 2014 schedule, which called for a first flight in 2022) with serial deliveries expected from 2028-2029. This schedule was based on the Russian Deputy Defence Ministers statement 'We are counting the first flight in the region of 2025-2026 and the beginning of mass production for 2028-2029' (Kret). In the first quarter of 2018 it was indicated that manufacture of the prototype PAK DA would commence at the Kazan Aviation Factory with a contract in place to build the aircraft (this did not indicate how many development aircraft would be built). By October 2018, the revised schedule called for the first flight in 2019 (brought forward some six years) and initial operations with service aircraft in 2025. Such an ambitious schedule was most certainly going to be subject to change.

By 2018, the PAK DA concept had further evolved into the design shown top and above, which, on the surface, appears to be smaller in overall dimensions than the flying wing concept of 2014 shown at the bottom of page 67. The design features six wheel main undercarriage units, a twin nose wheel unit, a flying wing layout with upturned winglets and two cockpit sections, indicating a two, three or four crew aircraft. Three sets of control surfaces are evident on each wing trailing edge and a large clamshell control surface is located on the rear centre section. TsAGI/MODRF

4

DISCUSSION

Russia's employment of its triad of strategic missile carrier/bombers over the Syrian Arab Republic in 2015-2017 demonstrated that nation's ability to project airborne strategic strike capability over vast distances. Air launched cruise missiles were launched from Tupolev Tu-95MS and Tu-160 strategic missile carriers and gravity bombs were dropped from Tupolev Tu-22M3 intermediate range bombers against ISIS (Islamic State) offensives, particularly in Deir ez Zor Province in Eastern Syria (Harkins, 2016).

Having a strategic missile carrier/bomber capability is hugely beneficial to the Russian Federation's defence strategy. However it has to be questioned, in light of the increased threat from a NATO (North Atlantic Treaty Organisation) military buildup on her borders and fostering of anti-Russian sentiment among her western and southern neighbours, whether the PAK DA program is affordable. The term affordable is tempered in the context of the costs being better applied in other areas. The stated United States of America intention to withdraw from the 1987 INF (Intermediate-range Nuclear Forces) treaty (the treaty came into effect on 30 June 1988) possesses unknowns for Russia's future defence planning. The INF treaty was without question pivotal in reducing East-West tension and accelerating arms reduction treaties that vastly reduced Soviet (later Russian Federation) and American nuclear arsenals that fostered a climate of mutual acceptance of coexistence in place of the confrontation of the Cold War decades. In the event that Europe, particularly East European nations, a number of which are openly hostile to the Russian Federation, host any potential American intermediate range missiles developed post INF treaty exit (in the event of the US withdrawal from the INF treaty such weapons would be expected to be developed as the only logical reason for withdrawal is not be bound by treaty obligations), then Russia would be forced to respond by developing and deploying intermediate range missiles as a countermove. The added expense of such a development/production/deployment would be an unwelcome burden on the Russian economy that may have to be offset by cancellation or reductions of one or more other defence program.

It would be folly for the Russian nation to allow herself to be dragged into the outdated concept of an arms race with the considerably economically stronger United States. The Russian Federation, unable to compete with the enormous military budget of the collective NATO she perceives as a direct threat to her borders may be forced to consider cancelling, or at least postponing projects such as the PAK DA. In its stead, a major investment could be placed in intermediate range hypersonic and aero-ballistic missile technology, which, unlike modern 'stealth' driven airborne strike assets, are immune to interception by current and projected air and missile defence systems. The advances in technology used to detect stealth aircraft dictates that these aircraft can be detected, tracked and engaged by modern air defence systems, albeit, at reduced effective ranges compared to non-stealth driven platforms. In this respect, a strategic bomber platform that relies on stealth technology as its main defence will come up wanting when faced with the most advanced air defence systems available. On the other hand, the introduction of PAK DA armed with long range hypersonic weapons will provide a dynamic complex that can be tasked for a number of mission scenarios. The types survivability will also be enhanced through introduction of advanced countermeasures systems and the stand-off capability of its onboard weapons – the Kh-101/102 air launched cruise missiles can be employed against targets ~4500 km from the launch point, allowing all NATO European (including the United Kingdom and Iceland), Mediterranean, Persian Gulf region and Arctic targets to be engaged when launched from a missile carrier flying within the protected airspace inside Russia's borders. Tu-160's operating from Anadyr in the Russian Far East could engage any target in Alaska, Japan, South Korea, much of Canada and targets in the North and much of the West United States without leaving Russian protected airspace. With a moderate Oceanic flight, the Tu-160/Kh-101/102 combination can strike targets as distant as the Hawaiian Islands and South West targets in the continental United States. When operating from Anadyr the Tu-22M3M, armed with the 2000 km range hypersonic Kinzhal, can strike targets in all of Alaska, Northern Canada, all of South Korea and most of Japan when the missiles are launched form within Russian protected airspace. Targets further afield can be struck when the launch platforms leave protected Russian airspace. On deep penetration nuclear strike missions a Tu-160 operating from Anadyr could target targets throughout the continental United States, but would have to leave protected airspace to achieve this. It is clear that the along with sea launched subsonic cruise missiles of the OKB Novator Kalibr-NK type, Long Range Aviation provides the Russian Federation with a significant intermediate range strike capability outside the constraints of the INF treaty which excluded sea launched and air launched missiles.

Russian defence spending is not too far ahead of that of the United Kingdom's, but falls to considerably less than one tenth of the collective NATO defence spending. If the oft quoted value in the region of US $68 billion is taken for Russian defence spending in 2017, this equates to ~13% of the US defence budget of more than $510 billion (2017). Despite inaccurate media reporting to the contrary, it is a fact that Russian defence spending has, in real terms, been falling over the several years to 2017, while that of collective NATO has accelerated. The reasons for the fall in Russian defence spending are twofold. 1. Much of the modernisation programs to bring the Russian Federation armed forces from Cold war era to modern equipment levels have been implemented or funded. 2. The Russian Federation government planning has turned to implementation of improvements in civil society. It is hard to shrug off the possibility, indeed, probability, that the reasoning behind NATO's increasingly threatening presence on Russian borders, vastly increased NATO military spending and the continuing imposition of economic sanctions on Russia are intended to strain her economy to the point that civil society improvements are cut back or abandoned in the hope that this will sow discord among the population – this may well be considered a war like situation whereby NATO hopes to inflict economic damage on Russia without resorting to an armed conflict.

While the increased threat posed by NATO withdrawal from arms control treaties and the NATO alliance support for governments hostile to Russian speaking peoples and the Russian Federation at large may force a further real terms increase in the Russian defence budget, this would be balanced with a review into the priority of some of the current and near term equipment programs. A Tu-160M2 armed with weapons in the class of the Kinzhal hypersonic cruise missile may be considered more survivable than a subsonic speed PAK DA, despite its low-observable qualities, armed with short-range nuclear tipped guided missiles or bombs as the latter aircraft design would have to penetrate defended airspace to deliver its payload.

Tu-95MS, Red 54, underwent the Stage 1 overhaul and modernization at the Beriev aircraft facility. MODRF/UAC

The Tu-160/hypersonic cruise missile (assuming a missile range of at least 2000 km) combination would. As noted above, be able to strike targets in most of Europe without leaving protected airspace, but would have to leave protected airspace to deliver missile strikes on the North American continent. The same combination, or the Tu-22M3M/hypersonic missile combination, would also be able to ensure NATO carrier strike groups are kept beyond the range required to strike Russian territory without extensive use of in-flight refueling assets. Such a combination would be equally capable of targeting NATO surface warships that ventured within BGM-109 Tomahawk cruise missile launch range of targets deep within Russian territory. This would, in effect, neutralise a major portion of NATO strike capability through asymmetric means at very small cost in comparison to the enormous sums spent on vast aircraft carrier and surface combatant naval groups.

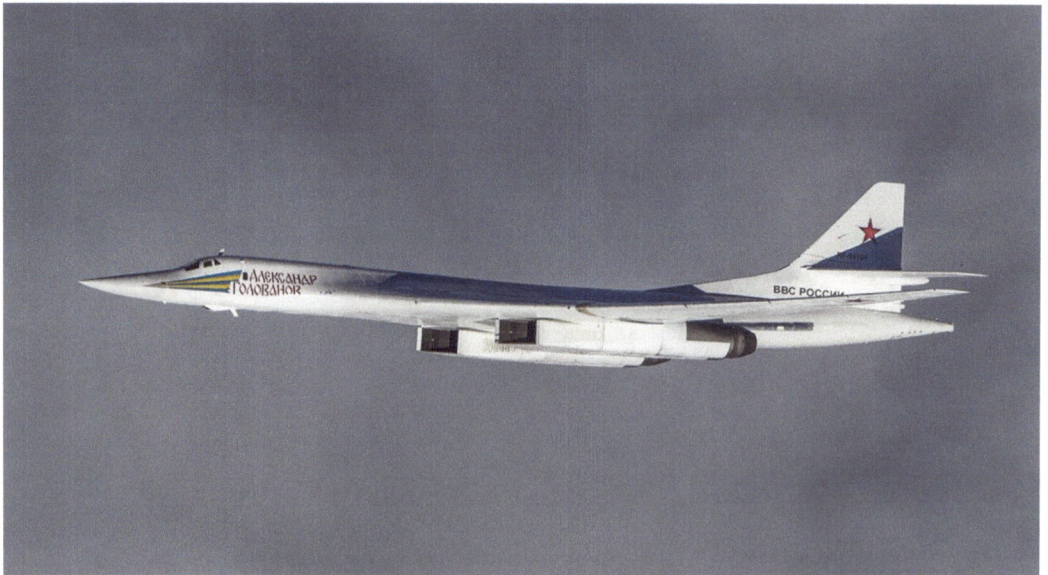

Top: A quartet of Long Range Aviation Tu-95MS strategic missile carriers at Engels air base, Saratov, prior to take-off for a training mission on 10 November 2018. Above: One of two Tu-160 strategic missile carriers on a flight that covered the skies over international waters of the Barents, Norwegian and North Seas on 21 November 2018. MODRF/UAC

APPENDICES

Appendix I

Tupolev Tu-95MS Strategic Missile Carrier – data furnished by UAC, PJSC Tupolev and MODRF

Engine: 4 x HK-12M (NK-12M) each generating 15,000 hp. for a combined power output of 60,000 hp.
Length: 49.13 m
Height: 13.3 m
Wing span: 50.4 m
Wing sweep position in the line of ¼ chord: 35°
Maximum take-off weight: ~187 tons
Maximum speed: 830 km/h
Service ceiling: 10500 m
Ordnance: Tu-95MS-6 can carry six Kh-55 or Kh-102 nuclear armed long-range cruise missiles or six conventional armed cruise missiles carried on a rotary launcher housed in the internal weapons bay. The Tu-95MS-16 can carry six cruise missiles on a rotary launcher in the internal weapons bay and 10 on under wing stations on the inner wing sections
Crew: Up to seven

Appendix II

Tupolev Tu-160 Strategic Missile Carrier – UAC, PJSC Tupolev and MODRF

Engines: 4 x HK-32 (NK-32) turbofan engines each rated at 25 tons (~22680 kg) in afterburner
Length: 54.10 m
Height: 13.2 m
Wing span: 55.7 m at 20° sweep, 50.7 m at 35° sweep and 35.6 m at maximum sweep back of 65°
Wing sweep position in the line of ¼ chord: 35°
Maximum take-off weight: 275000 kg (275 tons)
Maximum speed: 1800 km/h
Cruise speed: ~1030 km/h
Service ceiling: 14000 m
Range: ~14000 km maximum unrefueled or 10000 km with maximum ordnance
Ordnance: 22500 kg (Kret figures state 40000 kg maximum)
Crew: Four - pilot, co-pilot, navigator and systems operator

Appendix III

Tupolev Tu-22M3 Long-Range Missile Carrier/Bomber – UAC, PJSC Tupolev and
MODRF

Engines: Two HK-25 (NK-25) three-stage afterburning turbofans each rated at
25000 kgf (Height=0, Mach=0, ISA) with a specific fuel consumption at take-off of
2.08 kg/kgf/h
Length: 42.46 m
Height: 11.05 m
Wing span: 34.28 m at 20° sweep and 27.70 m at 65° sweep
Maximum take-off weight: 124 tons (124000 kg)
Normal combat load: 12000 kg
Maximum combat load: 24000 kg
Maximum speed: 2000 km/h
Cruising speed: 900 km/h
Service ceiling: 14000 m
Tactical range: 2200 km
Take-off run: 2000-2100 m
Landing roll: 1200-1300 m
Crew: Four - pilot, co-pilot, navigator and systems operator

Appendix IV

JSC Tactical Missiles Corporation (GosMKB Raduga,) Kh-101

Cruise speed: Unknown but thought to be similar to the Kh-55 ~850 km/h
Operational altitude: 30 m and upwards
Range: In excess of 4500 km (MODRF value)
Flight profile: dependent upon terrain to be overflown
Warhead type: Conventional
Warhead weight: thought to be 500 kg class
Accuracy: Within 5 m
Carrier: Tupolev Tu-95MS and Tu-160

Appendix V

2018 and Notional 2040 Strategic Missile Carrier/Bomber fleets		
Fleet	2018	2040
Tu-160	17[1]	
Tu-95MS-6	~29	
Tu-95MS-16	~35	
Tu-22M3	~70	20-40
Tu-160M2		30-50
Tu-95MSM		0-30
Tu-22M3M	1[2]	30
PAK DA[3]		

Note: The above values take no account of attrition through whatever means
Note 2: The 2040 notional fleet is based on very limited data released by various bureau and the MODRF and will certainly be subject to change. The value for the Tu-95MSM would vary considerably depending on the scale of production for the PAK DA

Appendix VI

Weapon	IOC	Range, km	Platform
Kh-22M	1970's	~600	Tu-22M3
Kh-55MS	1980's	~2000	Tu-95MS/Tu-160
Kh-101/102	2012	4500	Tu-95MS/Tu-160/M2
Kh-32	2019-20[4]	600+[5]	Tu-22M3M
Kinzhal	2017	2000	MiG-31K/Tu-22M3M (feasibility study)
XB-D			New Generation long-range missile

---

[1] Tu-160's are expected to be modernised to Tu-160M2 standard under 2018 planning

[2] Maiden flight conducted on 28 December 2018

[3] No values have been forwarded for planned PAK DA numbers

[4] Estimated. The complex is in trials operation in 2018

[5] This value is based on the ~600 km range of the Kh-22, but the Kh-32 may vary upwards or downwards

GLOSSARY

AI	Artificial Intelligence
ALCM	Air Launched Cruise Missile
B	Bomber
CIA	Central Intelligence Agency
CRRF	Collective Rapid Reaction Forces
CSTO	Collective Security Treaty Organisation (countries include – Russia Federation, Armenia, Belarus, Kazakhstan, Kyrgyzstan, Tajikistan and Uzbekistan)
DoD	Department of Defense
ECM	Electronic Counter Measures
EMP	Electro Magnetic Pulse
GLONASS	Globanaya Navigozionnaya Sputnikovaya Sistema (Global Navigation Satellite System
HF	High Frequency
Hp	Horsepower
Il	Ilyushin
IMA	Integrated Modular Avionics
INF	Intermediate-range Nuclear Forces
INS	Inertial Navigation System
IR	Infrared
ISIS	Islamic State of Iraq and the Levant
Kgf	Kilogram force
Km	Kilometer
km/h	Kilometers per hour
M	Metre
Mach	1 Mach = the speed of sound (this varies with altitude)
MAWS	Missile Approach Warning System
MiG	Mikoyan
MODRF	Ministry of Defence of the Russian Federation
NATO	North Atlantic Treaty Organisation
NAVSTAR	Navigation Satellite Timing and Ranging
PAK DA	Perspective Aviation Complex for Long Range Aviation
PJSC	Public Joint Stock Company
RAC	Russian Aircraft Corporation
RORSAT	Radar Ocean Reconnaissance Satellite
RWR	Radar Warning Receiver
TACAN	Tactical Navigation

TsAGI	Central Aerodynamic Institute
Tu	Tupolev
UAC	United Aircraft Corporation
UHF	Ultra-High Frequency
US	United States
USN	United States Navy
VKS	Russian Federation Air Force
=	Equals
°	Degree(s)
~	Approximately equal to (can also be used to mean asymptotically equal)

ABOUT THE AUTHOR

Hugh Harkins FRAS is a historian and author with an extensive research background in astro/geophysics and studies/research in the wider scientific, aeronautic, astronautic and nautical technical and historical fields. He is also involved in research in the field of Scottish history, which formed a significant element of an otherwise scientific undergraduate degree. Hugh has published in excess of sixty books; non-fiction and fiction, writing under his given name as well as utilising several pseudonyms. He has also written for several international magazines, whilst his work has been used as reference for many other projects ranging from the aviation industry, international news corporations and film media to encyclopaedias, museum exhibits and the computer gaming industry. Hugh is a member of the Institute of Physics and is an elected Fellow of the Royal Astronomical Society. He currently resides in his native Scotland. Other titles by the author include:

Russia's Coastal Missile Shield - Bal-E & Bastion Mobile Coastal Cruise Missile Complexes
Iskander - Mobile Tactical Aero-Ballistic/Cruise Missile Complex
Orbital/Fractional Orbit Bombardment System - The Soviet Globalnaya Raketa
Counter-Space Defence Co-Orbital Satellite Fighter
Sukhoi T-50/PAK FA - Russia's 5th Generation 'Stealth' Fighter
Sukhoi Su-35S 'Flanker' E - Russia's 4++ Generation Super-Manoeuvrability Fighter
Sukhoi Su-34 'Fullback'
Sukhoi Su-30MKK/MK2/M2 - Russo Kitashiy Striker from Amur
Soviet Mixed Power Experimental Fighter Aircraft – Piston-Liquid Propellant Rocket Engine/Piston-Ramjet/Piston-Pulsejet & Piston-Compressor Jet Engine Designs of the 1940's
MiG-35/D 'Fulcrum' F – Towards the Fifth Generation
Air War over Syria, Tu-160, Tu-95MS & Tu-22M3 - Cruise Missile and Bombing Strikes on Syria, November 2015-February 2016
Sukhoi Su-27SM(3)/SKM
Russian/Soviet Aircraft Carrier & Carrier Aviation Design & Evolution Volume 1 -
Seaplane Carriers, Project 71/72, Graf Zeppelin, Project 1123 ASW Cruiser & Project 1143-1143.4
Heavy Aircraft Carrying Cruiser
Light Battle Cruisers and the Second Battle of Heligoland Bight
British Battlecruisers of World War 1 - Operational Log, July 1914-June 1915
Eurofighter Typhoon - Storm over Europe
North American F-108 Rapier - Mach 3 Interceptor
Convair YB-60 - Fort Worth Overcast
Boeing X-36 Tailless Agility Flight Research Aircraft
X-32 - The Boeing Joint Strike Fighter
X-35 - Progenitor to the F-35 Lightning II
X-45 Uninhabited Combat Air Vehicle
Into The Cauldron - The Lancaster MK.I Daylight Raid on Augsburg
Hurricane IIB Combat Log - 151 Wing RAF, North Russia 1941
RAF Meteor Jet Fighters in World War II, an Operational Log
Typhoon IA/B Combat Log - Operation Jubilee, August 1942
Defiant MK.I Combat Log - Fighter Command, May-September 1940
Blenheim MK.IF Combat Log - Fighter Command Day Fighter Sweeps/Night Interceptions, September 1939 - June 1940
Fortress MK.I Combat Log - Bomber Command High Altitude Bombing Operations, July-September 1941

www.ingramcontent.com/pod-product-compliance
Lightning Source LLC
Chambersburg PA
CBHW041454210326
41599CB00005B/250